T0134922

Studies in Computational Intelligence

Volume 824

The series "Studies in Computational Intelligence" (SCI) publishes new developments and advances in the various areas of computational intelligence—quickly and with a high quality. The intent is to cover the theory, applications, and design methods of computational intelligence, as embedded in the fields of engineering, computer science, physics and life sciences, as well as the methodologies behind them. The series contains monographs, lecture notes and edited volumes in computational intelligence spanning the areas of neural networks, connectionist systems, genetic algorithms, evolutionary computation, artificial intelligence, cellular automata, self-organizing systems, soft computing, fuzzy systems, and hybrid intelligent systems. Of particular value to both the contributors and the readership are the short publication timeframe and the world-wide distribution, which enable both wide and rapid dissemination of research output.

The books of this series are submitted to indexing to Web of Science, EI-Compendex, DBLP, SCOPUS, Google Scholar and Springerlink.

More information about this series at http://www.springer.com/series/7092

Lorenzo Cevallos-Torres ·
Miguel Botto-Tobar

Problem-Based Learning: A Didactic Strategy in the Teaching of System Simulation

Lorenzo Cevallos-Torres
Faculty of Mathematical and Physical Sciences
University of Guayaquil
Guayaquil, Ecuador

Miguel Botto-Tobar
Eindhoven University of Technology
Eindhoven, The Netherlands
University of Guayaquil
Guayaquil, Ecuador

ISSN 1860-949X ISSN 1860-9503 (electronic)
Studies in Computational Intelligence
ISBN 978-3-030-13395-5 ISBN 978-3-030-13393-1 (eBook)
https://doi.org/10.1007/978-3-030-13393-1

Library of Congress Control Number: 2019931854

This Springer imprint is published by the registered company Springer Nature Switzerland AG
The registered company address is: Gewerbestrasse 11, 6330 Cham, Switzerland

We dedicate this book to our Family.

Preface

This book is designed to be a general, descriptive, and didactic introduction about the use of systems simulation with probability distributions, in this way might become a tool that helps to solve real-life problems, by evaluating alternative scenarios and finding answers to questions like “what would happen if?”.

This book aims the application of systems simulation as the main contribution in the teaching-learning model. We will analyze the different types of methodologies used in teaching to solve real-life problems, followed by process of building a simulation model by using a computer and probability distributions, allowing the correlation between a real model and a simulated model.

The conceptual part aims that the different methods are understood as a research instrument subjected to continuous review that allows a progressive refinement in the compression of system, which leads to a suitable position to make decisions in the problem solutions.

The systems’ simulation is based on the concept of experimentation itself of the scientific method, according to which, the experiments are carried out on a dynamic model instead of the real system so that the model results might be a valid representation of the system.

Consequently, one goal of this book is to help understand how the different simulation methods might be used to analyze phenomena and problems, and make decisions about them, i.e., to demonstrate the role of simulation in the processes of decision-making, especially in computer systems with a pedagogical teaching-learning approach.

The simulation allows getting better analysis and evaluation of the system’s performance before they are built. Thus it becomes a vital design tool, in any of its phases, and moreover, to estimate a priori the impact of proposed changes in the existent systems. It is expected to illustrate how the simulation can be applied in wide situations, through small projects such as: transport system analyze, logistics, queue’s theory, inventory’s theory, medicine.

Finally, to help the reader not only to the conceptual understanding of different methods of systems simulation, but also to understand how they work; when they should be used, and when not; what can expect from the simulation; what errors must be avoided in the development and use; and how the simulation might help to improve the performance of systems.

Guayaquil, Ecuador
December 2018

Lorenzo Cevallos-Torres
Miguel Botto-Tobar

Acknowledgements

We would like to thank God Almighty for giving us the strength, knowledge, ability, and opportunity to undertake this research study, and also to who are the invaluable protagonists who put their grain of sand in the elaboration of this book, our students at the University of Guayaquil.

Contents

About the Authors

Lorenzo Cevallos-Torres received his Bachelor in Computer Statistics Engineering, at Escuela Superior Politéctina del Litoral, Ecuador in 2005. He then received an M.Sc. degree in Productivity and Quality Management at the Escuela Superior Politéctina del Litoral, Ecuador in 2015, and an M.Sc. degree in Computational Modeling in Engineering at the Universidad de Cádiz, España in 2015. He joined the Department of Mathematics, and Physic Sciences at the University of Guayaquil, Ecuador as Assistant Professor in 2012–2015, as Professor from 2016. He is the Editor of the books "Análisis Estadístico Univariado" published by the Editorial University of Guayaquil and "El uso de las tecnologías de la información y la comunicación como estrategia de enseñanza-aprendizaje de la Estadística Univariada" published by Editorial Académica Universitaria, Universidad de las Tunas Cuba. His current research areas include Probability, Statistics, Quality management, and Soft computing.
University of Guayaquil, Ecuador,
lorenzo.cevallost@ug.edu.ec

Miguel Botto-Tobar received his Bachelor in Computer System Engineering at Universidad Estatal de Milagro, Ecuador in 2010. He then received an M.Sc. degree in Software Engineering, Formal Methods and Information Systems at the Polytechnic University of Valencia, Spain in 2014. Currently, he is doing a Ph.D. in Computer Science at the Eindhoven University of Technology in the Netherlands. He joined the Department of Mathematics, and Physic Sciences at the University of Guayaquil, Ecuador as Assistant Professor in 2016. He is a recipient of the "Convocatoria Abierta" Scholarships (2011–2013) of the Ecuadorian government. He is the Editor of the books "Technology Trends" and Co-Editor "Information and Communication Technologies of Ecuador (TIC.EC)" all published by Springer, Switzerland. He is an Associate Editor of the Journal of Science and Research, and the Neutrosophic Computing and Machine Learning Journal. His current research areas include Software Engineering, Empirical Software Engineering, Model-Driven Development, and Human Aspects.

Eindhoven University of Technology, the Netherlands m.a.botto.tobar@tue.nl, and University of Guayaquil, Ecuador, miguel.bottot@ug.edu.ec

Chapter 1
The System Simulation and Their Learning Processes

In the 21st century, Education is undergoing a series of transformations both inside and outside the classroom. Despite changes in education, knowing and understanding the teaching-learning process is key to creating effective pedagogical action. In the process, the most significant teacher task is to accompany the student's learning. Being this accompaniment through a concise and legible exhibition of concepts and projects performed in the classroom; that is to say, a way of driving theory into practice the basic processes around the systems simulation as well as their implications regarding analysis, and implementation of information technology.

1.1 Introduction

To determine a way to carry the students training process at the university level. It is necessary to consider the appropriate use of teaching tools that will be used to complete the different formative stages of this student. A project-based learning methodology (PBL) has been used where the student develops competencies in a collaborative approach in search of real solutions [1–3].

Simulation is a way of approaching the study of any real dynamic system where it is feasible to have a behavior model, and in which the variables and parameters that characterize it can be distinguished. In order to make this possible, we proceed to the use of mathematical and probabilistic tools [4, 5].

One of the difficulties faced by the university student is to understand the proper use of the learning of the simulation subject. For this, the student must possess very marked skills within the teaching-learning process, as it is, the domain of the previous mathematical and statistical problems [6–9].

In this book, we expose the experiences achieved of applying the PBL methodology in teaching of "systems simulation" subject. The most significant result is the evidence that the use of this educational alternative favors the students' motivation

L. Cevallos-Torres and M. Botto-Tobar, *Problem-Based Learning: A Didactic Strategy in the Teaching of System Simulation*, Studies in Computational Intelligence 824, https://doi.org/10.1007/978-3-030-13393-1_1

for the subject insofar as it manages to relate it to their profession and recognize its importance in solving problems of society and the environment where they coexist daily [6, 10–12].

1.2 Fundamentals of Simulation

In the first analysis of these definitions, it reveals the relationship between simulation as a teaching method and modeling as a general scientific method of obtaining knowledge. Through simulation, the student will not work directly with the study object, but with a representation of such an object, from which the most important elements are abstracted, taking into account the purposes pursued. This invariable situation means modeling [13–15].

There are several simulation modalities: experimental, methodological, instrumental and decision-making. This last variant is based on the fact that the student must develop the exercise by making decisions to reach a final result and thereby determine the path to follow in the process. The use of simulation makes it possible to accelerate the learning process and contributes to raising its quality [16–18].

The importance of simulation as a method is that it reproduces real objects when, due to problems of time, resources or security, it is not possible to carry out the activity in its natural environment, with its true components. Hence, the wide use of this method arises, since it is practically applied in all disciplines and branches of science. Similarly, modeling facilitates the analysis of the original processes in those cases in which it is expensive, difficult or impossible to investigate real objects [19, 20].

1.3 Simulation as a Teaching Method and Its Link with Engineering Careers

In the University of Guayaquil context, the system simulation is a fundamental subject in the curriculum in the computer systems engineering career. It is studied in the sixth semester, and it provides the knowledge related to the application of real-life problems, simulated in a computer environment, which allows the student to develop specific strategies that help him/her to efficient learning and the capacity of self-learning [21–23].

In Project-Based Learning (PBL), the university teacher in charge of the group of students of the system simulation course acts as a tutor instead of being a conventional master expert in the area and a knowledge transmitter; which will help the students to reflect, identify the needed information and motivate them to continue with the work, in the other words, they will guide them to reach the proposed learning goals [24–26].

In order to achieve the success of the PBL methodology in the classroom; the tutor should not be considered as a simple passive observer, on the contrary, he/she should be active in guiding the learning process making sure that the group does not lose the goal set. And also identifies the most important issues to meet the resolution of the problem, making the student progress appropriately towards the achievement of learning objectives, in addition to identify what they need to study for a better understanding, meaning, what they learn in theory to apply it in practice based on learning real-life problems [27–29].

The tutor who is in charge of the subject supports the development of the skills that the student needs to get ahead, so it can assure that the tutor turns out to be a fundamental piece for the development of the PBL methodology, in fact, the dynamics of the work process of the group depends on its good performance [30–32].

In order to understand the use of simulation as a base tool in the educational process; it has been considered to analyze it through two major uses:

- During teaching-learning.
- In the evaluation.

During teaching-learning, the use of the system simulation allows to the student modeling different work environments, so that, through a small data collection in the experimentation, the student will improve their learning process to handle diagnostic, treatment and problem-solving techniques properly. It will also help them to improve their cognitive and critical thinking skills. The simulation allows learners to focus on a specific teaching goal; it enables the reproduction of a particular procedure or technique and to apply a standardized criterion [33, 34].

It is important to be clear that simulation imitates but does not reproduce what happens in a real environment exactly, and we also consider certain aspects regarding the use of simulation models that might be presented [35].

- There are aspects of reality that cannot be simulated, a matter that must be kept in mind whenever we use any simulation.
- It is necessary to be very cautious in predicting; based on the response to a simulated situation and how a person will be led in the face of real situations.
- We cannot restrict the development of skills nor the evaluation of students' performance only through simulation, therefore, combining the use of different methods and resources in order to obtain consistent and closely results to real situations [36].

It is important to be clear that the use of systems simulation has two main pillars of teaching and learning such as mathematical modeling and the proper use of high-generation computer programs. The student in the systems engineering career has many skills in computing management and advanced computer packages, otherwise, when the student has to interpret the information mathematically, it is complicated, since he has no skills in the management of mathematical modeling [37–39].

1.4 Teaching-Learning Process

The teaching-learning process is conceived as the space in which the main protagonist is the student, and the teacher fulfills a function as a facilitator. It is also considered as a procedure, through which special or general knowledge about a particular subject is transmitted, its dimensions in the phenomenon of academic performance from the factors that determine its behavior in the classroom. Teachers in the search for solutions to routine situations are concerned with developing a particular type of motivation in their students, which encourages their interest in learning, which consists of many elements [40, 41].

The collaborative interaction between teachers and students and/or between themselves determines that learning can be evidenced through a test or evaluation to students of a given semester or student cycle. The learning must be evidenced since to be considered as reached it has to maintain a sufficient duration. Besides, the reflection of motivation, behavior, social relationships and other factors converges [42, 43].

Learning is a relatively permanent change in the behavior of an individual that reflects an increase in knowledge and expands the area of potential development, also strengthens the intelligence or skills achieved that are reflected in practice [44].

1.5 Project Based Learning (PBL)

Project Based Learning (PBL) can be defined as a teaching and learning modality focused on real-life tasks and projects, whose main objective is to obtain a product that helps the student to understand with clarity what they have learned, theoretically in the classroom. It is important to indicate that this method promotes individual and autonomous learning within a work plan defined by objectives and procedures, that is, to determine what extent the student achieves the goals set out in advance; to achieve it important that students take responsibility for their learning, discover their preferences and strategies in the process [11, 21, 27].

An advantage of PBL is that the student acquires throughout the teaching process, skills and competencies such as collaboration, project planning, communication, decision making, and time management. In other words, PBL is a learning model in which students pose, implement and evaluate projects that have real-world application [10, 15, 34].

It is important to be clear that in the PBL model, interdisciplinary, long-term and student-centered learning activities are developed beyond the classroom [45]. It motivates the students to learn, and it particularly gives them the option of selecting subjects that interest them and that are important for their professional life, by making that connection between the learning they acquire in the university in a theoretical and not very comprehensible way, combined with the reality of the environment where we live, and it makes students retain more knowledge and skills, this is as long as the student is really committed to the project development, which at one point throughout their learning process become stimulants [46–48].

In seeking to connect the university with the real world [49], the teacher must adopt a different role to the one he has been carrying for a long time, that is, the teacher must change the way he has been teaching traditionally [50]. From this point the teacher stops being a simple transmitter of knowledge, and must become a planner, a facilitator [26], since, his/her function will be planning, observing, accompanying, stimulating and evaluating the learning situations, that is, the teacher role focuses on [51]:

- Prepare the learning process meticulously.
- Keep themselves in the second plane as much as possible taking note of what works and what does not.
- Be available to answer questions and solve concerns from students.
- Encourage students to learn themselves and ask the right questions.
- Encourage students to self-assess their work and experiences.
- Remember what is forgotten and overlooked, but it should be considered and developed, together with students, the needed content based on practical experience.
- Pay special attention to cooperation aspects, tasks organization, group work methodology, and include them in conversations with other students.
- Evaluate.

The PBL is a learning process centered on the student, and it hopes a behavior series and different participation to those required in the conventional learning process. Some desirable characteristics in the students who participate in the PBL are presented below. Besides, it is important to point out that if the students do not have these qualities, they must be willing to develop or improve them. Deep and clear motivation about the needed learning [12, 34, 38, 42, 52].

- Disposition to work in a group.
- Tolerance to face ambiguous situations.
- Skills for personal interaction both intellectual and emotional.
- Development of the imaginative and intellectual powers.
- Skills for problem-solving.
- Communication skills.
- See your study field from a broader perspective.
- Critical, reflective, imaginative and sensitive thinking skills.
- Target content for teaching and learning.

The PBL is one of the active teaching-learning methods that has had the most significant impact on higher education in recent years. Due to the path is taken by the conventional process is reverted [11, 13].

The goal of this study is to present the implementation results of the PBL method throughout the simulation course in the computer systems engineering career at the UG [21].

Based on our experience, the PBL is students-centered and highly effective method to stimulate their activities and educate their creative scientific thinking. For its application, we organize students in groups and select experiment as a teaching activity [27].

Regarding the student learning process, it is important to recognize the existence of several tools, from a computational scope as general purpose computing resources or general-purpose programs, which are the computer applications that can be useful for all types of computer users [13].

1.6 Role of the Teacher in the Teaching Process

This learning model requires teachers to adopt a different role than usual in traditional teaching. They stop being mere "transmitters of knowledge" becoming facilitators and planners of learning their function is to plan, observe, accompany, stimulate and evaluate learning situations.

1.6.1 Observe

At this point, the teacher tends to observe processes, changes, behaviors, relationships, difficulties, and potentials that can help students to develop the process, to establish changes or improvements and reach the achievement of the objectives [53–55].

1.6.2 Accompany

In the projects, the faculty assumes the role of accompanist or mediator and provides support regarding content and method. In any case, there may be a "client" who is responsible for the project, but at other times, the same teacher will have to say what is required both at the beginning and throughout the project. Also, it takes into account the fit with related training contents and with others of a professional and systematic nature [56–58].

1.6.3 Stimulate

Project development work offers a lot of room to make decisions of your own and to develop creative possibilities. The key is to trust the students in such a way that they can improve themselves, as a team, a plan of solutions to the problem and solve it. It is important that the teacher is clear about what he/she is trying to achieve and, thus, helps the student to discover what these objectives are to capture their interest [59–61].

1.6.4 Evaluate

There must be a reward for the student's and the group's achievements, in the beginning, during and at the end of the learning. The evaluation is a process that accompanies the entire training project. Working for projects changes the relationships between teachers and students. It reduces competition among students and allows students to collaborate and work with each other. The contributions of others are accepted as help and not as competition. Also, projects can change the approach to learning: from the simple memorization of facts to the exploration of ideas and the development of skills and tools [62–64].

1.7 The Use of ICT in the Teaching Process

Information and communication technologies currently have a great influence on university education, as well as within the academic curriculum of the curriculum of the UG computer systems engineering career; the simulation program has acquired an important degree of importance in recent times because these programs help solve real-life problems through the use of systems simulation. It is important to keep in mind that there is no ideal teaching method, so it is necessary to determine a selection of suitable applications that depend on the existing conditions for learning. The method used must correspond to the scientific level of the content, which will stimulate the creative activity and motivate the development of cognitive interests that link the school with life. It must, therefore, break the scholastic, rigid, traditional schemes and promote the systematization of the learner's learning, bringing it closer and preparing it for its work in society [2, 24, 26].

The simulation techniques are developed by looking for mathematical algorithms that speed up the calculations and especially incorporate the object-oriented technologies of the modern programming languages to create flexible, expandable and exportable code prototypes. The development of computer technologies will undoubtedly facilitate the opening of new projects; it may never be possible to define a complete computer learning system given that the basic technologies are constantly advancing [65, 66].

The use of simulation in educational processes in the computer systems engineering career at the University of Guayaquil is an effective teaching and learning method to achieve in our students the development of a set of skills that make it possible to achieve superior modes of action. As a result of the work carried out, it was possible to obtain the data corresponding to the unknowns raised and determined that: as for the analysis of the number of people who arrived at the IESS stop between 6:00 pm and 7:00 pm during the five days during the week. It was possible to determine that the average of people per day was around 2500 to 3000 people. The Poisson distribution was applied, which allowed generating the following graphs using Monte Carlo to increase the base data [6, 33, 67].

The purpose of system simulation is to offer to the sixth-semester students the opportunity to practice in the classroom, in such a way that there is an interaction with reality in the different areas in question; being these in the inventories area, logistics, and queuing theory. The simulation consists in situating the student in a context that imitates some aspect of reality and in establishing in that environment, problematic or reproductive situations, similar to those that he would have to face in real situations [37, 38, 67].

References

1. de Miguel Díaz, Mario. 2005. Cambio de paradigma metodológico en la educación superior. exigencias que conlleva. *Cuadernos de integración europea* 2: 16–27.
2. Rosales, M. 2014. *Proceso evaluativo: evaluación sumativa, evaluación formativa y assesment su impacto en la educación actual*, 4. In Congreso Iberoamericano de Ciencia, Tecnología: Innovación y Educación, vol.
3. Cabero Almenara, Julio. 2014. Formación del profesorado universitario en tic. aplicación del método delphi para la selección de los contenidos formativos. *Educación*, XX1, 17 (1): 111–132.
4. Olivos, Moreno, and Tiburcio. 2011. Didáctica de la educación superior: nuevos desafíos en el siglo xxi. *Perspectiva educacional* 50 (2): 26–54.
5. Maldonado, Carlos Eduardo, and Nelson Alfonso Gómez-Cruz, et al. 2010. Modelamiento y simulación de sistemas complejos. *Borradores de Investigación: Serie documentos Administración*. ISSN 0124-8219, No. 66 (Febrero de 2010).
6. Goñi, Joaquín, and Juan Martín García. 2006. Dinámica de los sistemas biológicos modelando-complejidad. *Inicialización a la Investigación. RevistaElectrónica* 1: 1–9.
7. Barneto, Agustín García, and Mario Rafael Gil Martín. 2006. Entornos constructivistas de aprendizaje basados en simulaciones informáticas. *Revista electrónica de Enseñanza de las Ciencias* 5: 1–19.
8. Mestras, Juan Pavón, Adolfo López Paredes, José Manuel Galán Ordax, et al. 2012. Modelado basado en agentes para el estudio de sistemas complejos. *Novática* 218: 13–18.
9. Fredes, Claudio A., Juan P. Hernández, and Daniel A. Díaz. 2012. Potencial y problemas de la simulación en ambientes virtuales para el aprendizaje. *Formación universitaria* 5 (1): 45–56.
10. Ferrer, David Macías. 2007. Las nuevas tecnologías y el aprendizaje de las matemáticas. *Revista Iberoamericana de Educación* 42 (4): 2.
11. Trasobares, Alejandro Hernández, and Gilaberte, Raquel Lacuesta. 2007. Aplicación del aprendizaje basado en problemas (pbl) bajo un enfoque multidisciplinar: una experiencia práctica. In *Conocimiento, innovación y emprendedores: camino al futuro*, 3. Universidad de La Rioja.
12. Pérez, Marisabel Maldonado. 2008. Aprendizaje basado en proyectos colaborativos. una experiencia en educación superior. *Laurus*, 14 (28): 158–180.
13. Fernández, Flavio H., and Julio E. Duarte. 2013. El aprendizaje basado en problemas como estrategia para el desarrollo de competencias específicas en estudiantes de ingeniería. *Formación universitaria* 6 (5): 29–38.
14. García-Almiñana, Daniel, and Beatriz Amante García. 2006. *Algunas experiencias de aplicación del aprendizaje cooperativo y del aprendizaje basado en proyectos*. In I Jornadas de Innovación Educativa: Escuela Politécnica Superior de Zamora.
15. Izquierdo, Luis R., José Manuel Galán Ordax, José I. Santos, Ricardo Del Olmo, and Martínez. 2008. Modelado de sistemas complejos mediante simulación basada en agentes y mediante dinámica de sistemas. Empiria. *Revista de metodología de ciencias sociales* 16: 85–112.
16. Mariño, Sonia Itatí, and María Victoria López. 2008. Un proyecto de docencia, extensión e investigación en la asignatura modelos y simulación. In *X Workshop de Investigadores en Ciencias de la Computación*.

17. Montenegro Marín, Carlos Enrique, Paulo Alonso Gaona García, JUAN CUEVA LOVELLE, and Óscar Sanjuán Martínez. 2011. Aplicación de ingeniería dirigida por modelos (mda), para la construcción de una herramienta de modelado de dominio específico (dsm) y la creación de módulos en sistemas de gestión de aprendizaje (lms) independientes de la plataforma. *Dyna*, 78 (169).
18. Giraldo, Jaime A., Carlos A. Toro, and Fabián A. Jaramillo. 2013. Aprendiendo sobre la secuenciación de trabajos en un job shop mediante el uso de simulación. *Formación universitaria* 6 (4): 27–38.
19. Rodríguez, José Bravo. 2000. *Planificación del diseño en entornos de simulación para el aprendizaje a distancia*. No. 108. Univ de Castilla La Mancha.
20. Primorac, Carlos R., Sonia Itatí Mariño, and María Victoria López. 2010. Simuladores para afianzar conceptos de modelos de existencias. In *V Congreso de Tecnologías en Educación y Educación en Tecnonolgías*.
21. Embuz, E., and J.D. Fernández-Ledesma. 2015. Propuesta de un método para la aplicación de un modelo de simulación basada en agentes del sistema regional de innovación. *Investigación e Innovación en Ingenierías* 3 (2).
22. Shoikova, Elena, and Slavka Tzanova. 1999. Innovaciones en la educación superior en electrónica a través del desarrollo de un entorno de aprendizaje basado en simulaciones y conducido por desarrollo de proyectos. *RIED. Revista iberoamericana de educación a distancia*, 2 (2): 125–135.
23. Viudes, María Victoria López, Sonia Itati Mariño Fernández, and Jaquelina Edit Escalante Saiach. Evaluar para integrar los contenidos en un curso: el caso de la asignatura modelos y simulación/evaluate to integrate contents in a course: subject models and simulation case. *Actualidades Investigativas en Educación*, 9(1).
24. Rodríguez, Ruth, and Rafael Bourguet. 2014. Diseño interdisciplinario de modelación dinámica usando ecuaciones diferenciales y simulación. In *Proceedings of the LACCEI Latin American and Caribbean Conference for Engineering and Technology (LACCEI'2014) "Excellence in Engineering To Enhance a Country's Productivity 22–24 de julio de 2014*.
25. Yániz, Concepción. 2008. Las competencias en currículo universitario: implicaciones para diseñar el aprendizaje y para la formación del profesorado. *Revista de docencia*. universitaria.
26. Salinas, Jesús. 2004. Innovación docente y uso de las tic en la enseñanza universitaria. *International Journal of Educational Technology in Higher Education (ETHE)*, 1 (1).
27. Bozu, Zoia, and Pedro José Canto. 2009. El profesorado universitario en la sociedad del conocimiento: competencias profesionales docentes. *Revista de formación e innovación educativa universitaria* 2 (2): 87–97.
28. Vanesa, Ausín, Víctor Abella, Vanesa Delgado, and David Hortigüela. 2016. Aprendizaje basado en proyectos a través de las tic: Una experiencia de innovación docente desde las aulas universitarias. *Formación universitaria* 9 (3): 31–38.
29. Navarro, Leonor Prieto. 2006. Aprendizaje activo en el aula universitaria: el caso del aprendizaje basado en problemas. *Miscelánea Comillas. Revista de Ciencias Humanas y Sociales*, 64 (124): 173–196.
30. Gessa Perera, Ana. 2011. La coevaluación como metodología complementaria de la evaluación del aprendizaje: análisis y reflexión en las aulas universitarias.
31. Raquel, Lacuesta, and Carlos Catalán. 2004. Aprendizaje basado en problemas: una experiencia interdisciplinar en ingeniería técnica en informática de gestión. *X Jornadas de Enseñanza Universitaria de la Informática* 305–311.
32. Martínez, Javier. 2004. El papel del tutor en el aprendizaje virtual. *línea].[consultado 22 julio 2004]. Disponible en Internet*: (Gerente de Desarrollo de Proyectos de GEC Chile). http://www.uoc. edu/dt/20383/
33. González, Manuel Álvarez. 2008. La tutoría académica en el espacio europeo de la educación superior. *Revista interuniversitaria de formación del profesorado* 61: 71–88.
34. Perea, Salas, S. Ramón, and Plácido Ardanza Zulueta. 1995. La simulación como método de enseñanza y aprendizaje. *Educación Médica Superior* 9 (1): 3–4.

35. Durán, Elena B., and Rosanna N. Costaguta. 2008. Experiencia de enseñanza adaptada al estilo de aprendizaje de los estudiantes en un curso de simulación. *Formación universitaria* 1 (1): 19–28.
36. Contreras, Gloria, Rosa García Torres, and María Soledad Ramírez Montoya. 2010. Uso de simuladores como recurso digital para la transferencia de conocimiento. *Apertura: Revista de Innovación Educativa*, 2 (1): 86–100.
37. Gómez, B., and Luz María. 2004. Entrenamiento basado en la simulación, una herramienta de enseñanza y aprendizaje. *Revista colombiana de anestesiología*, 32 (3).
38. Gordillo, Mariano Martín. 2003. Metáforas y simulaciones: alternativas para la didáctica y la enseñanza de las ciencias. *Revista Electrónica de Enseñanza de las Ciencias* 2 (3): 377–398.
39. Gisbert Cervera, Mercè, José Maria Cela, and Sofia Isus. 2010. Las simulaciones en entornos tic como herramienta para la formación en competencias transversales de los estudiantes universitarios. *Teoría de la educación: educación y cultura en la sociedad de la información, 2010*, vol. 11, núm. 1, 352–370.
40. Mariño, Sonia Itatí, and María Victoria López. 2007. Aplicación del modelo b-learning en la asignatura "modelos y simulación" de las carreras de sistemas de la facena-unne. *Edutec. Revista Electrónica de Tecnología Educativa* 23: 075.
41. Joyce, Bruce R., Marsha Weil, and Emily Calhoun. 2002. Modelos de enseñanza.
42. Ruiz, Ascensión Palomares. 2011. El modelo docente universitario y el uso de nuevas metodologías en la enseñanza, aprendizaje y evaluación the educational model at university and the use of new methodologies for teaching, learning and assessment. *Revista de educación* 355: 591–604.
43. Domingo, J. Contreras, and ENTRE PENSAMIENTO Y ACCION LA TENSION. 1987. De estudiante a profesor: socialización y aprendizaje en las prácticas de enseñanza. *Revista de educación no 282. Teoría del currículo*, 203.
44. ACEVEDO, PEDRO AHUMADA., 2001. *La evaluación en una concepción de aprendizaje significativo*. Chile: Ediciones Universitarias de Valparaíso.
45. Rojo, Fernando Lara. 2003. y de adquisición de conocimiento.
46. Blank, William E., and Sandra Harwell. 1997. Promising practices for connecting high school to the real world.
47. Maqueo, Ana María. 2006. *Lengua, aprendizaje y enseñanza: el enfoque comunicativo: de la teoría a la práctica*. Editorial Limusa.
48. Galagovsky, Lydia R. 2004. Del aprendizaje significativo al aprendizaje sustentable: parte 1, el modelo teórico. *Enseñanza de las Ciencias* 22 (2): 229–240.
49. Paixão, M., and António Cachapuz. 1999. La enseñanza de las ciencias y la formación de profesores de enseñanza primaria para la reforma curricular: de la teoría a la práctica. *Enseñanza de las Ciencias* 17 (1): 069–77.
50. Norman, Fairclough. 2008. El análisis crítico del discurso y la mercantilización del discurso público: las universidades. *Discurso & Sociedad* 2 (1): 170–185.
51. Fernández Muñoz, Ricardo. 2001. El profesor en la sociedad de la información y la comunicación: nuevas necesidades en la formación del profesorado.
52. Jiménez Pérez, Roque, and Ana Ma Wamba Aguado. 2004. ¿ podemos construir un modelo de profesor que sirva de referencia para la formación de profesores en didáctica de las ciencias experimentales? Profesorado. *Revista de Currículum y Formación de Profesorado* 8 (1).
53. Díaz, Espinosa, and Perla, and Carlos Villarroel González. 2007. Proposición y simulación de un algoritmo adaptativo para sistemas de antenas inteligentes. *Ingeniare. Revista chilena de ingeniería* 15 (3): 344–350.
54. Porlán Ariza, Rafael et al. 2011. El maestro como investigador en el aula: investigar para conocer, conocer para enseñar.
55. Jiménez, Esther Prieto. 2008. El papel del profesorado en la actualidad. su función docente y social. *Foro de educación*, 6 (10): 325–345.
56. de Medrano Ureta, Consuelo Vélaz. 2009. Competencias del profesor-mentor para el acompañamiento al profesorado principiante. *Profesorado. Revista de curriculum y formación de profesorado*, 13 (1): 209–229.

57. Ibeth, Gladys, and Ordóñez, and Ariza, and Héctor Balmes Ocampo Villegas. 2005. El acompañamiento tutorial como estrategia de la formación personal y profesional: un estudio basado en la experiencia en una institución de educación superior. *Universitas Psychologica* 4 (1): 31–42.
58. Lebrija, Analinnette, Rosa del Carmen, and Flores, and Mayra Trejos. 2010. El papel del maestro, el papel del alumno: un estudio sobre las creencias e implicaciones en la docencia de los profesores de matemáticas en panamá. *Educación matemática* 22 (1): 31–55.
59. Gavilán Izquierdo, José María. 2005. El papel del profesor en la enseñanza de la derivada. análisis desde una perspectiva cognitiva.
60. Maureira T., et al. 2008. Estrategia de acompañamiento a establecimientos educacionales vulnerables.
61. Vezub, Lea F. Teacher training and professional development in light of the new challenges of the school system. *Profesorado, Revista de Currículum y Formación del Profesorado*, 11 (1): 23.
62. Vallejo, Pedro Morales. 2006. Implicaciones para el profesor de una enseñanza centrada en el alumno. *Miscelánea Comillas. Revista de Ciencias Humanas y Sociales*, 64 (124): 11–38.
63. Álvarez Méndez, Juan Manuel. 2001. *Evaluar para conocer, examinar para excluir*. Morata.
64. Álvarez Méndez, Juan Manuel. 2008. Evaluar el aprendizaje en una enseñanza centrada en competencias. *Educar por competencias,¿ qué hay de nuevo?*, 206.
65. Hernández Díaz, Adela. 2002. Las estrategias de aprendizaje como un medio de apoyo en el proceso de asimilación. *Revista Cubana de Educación Superior*.
66. Gil, Jesús Vidal. 2006. *Un método general, sencillo y eficiente, para la definición y simulación numérica de sistemas multicuerpo*. PhD thesis, Universidad Politécnica de Madrid, 2006.
67. Martín, José Juan Barba, Roberto Monjas Aguado, Jesús Gómez García, Esther Matilde López Pastor, Juan F Martín Pinela, Javier González Badiola, Juan Carlos Manrique Arribas, Rebeca Aguilar Baeza, Marta González Pascual, Carlos Heras Bernardino, et al. 2006. La evaluación en educación física: revisión de modelos tradicionales y planteamiento de una alternativa: la evaluación formativa y compartida. *Retos: nuevas tendencias en educación física, deporte y recreación*, (10) : 31–41.

Chapter 2
Process Sampling

This chapter provides a basic introduction to descriptive statistic including sampling techniques such as population, simple random sampling, systematic random sampling, stratified sampling, cluster sampling, and non-probabilistic sampling to have enough knowledge to be able to decide which is the most appropriate sampling technique.

2.1 Population

A population is defined as a finite or infinite set of people or objects that have common characteristics. In other words, it is this whole phenomenon to be studied, where the population units have a common feature, which is studied and gives rise to the research data [1–3].

In Statistics, the elements of a population are defined as those individual units that constitute a population. It is also defined as any complete group, either people, animals or things. It is the whole set under consideration, and it refers to a finite group of elements (see Fig. 2.1) [4, 5].

2.1.1 Population Types

According to the number of individuals that conform the statistical population [7–9], it could be classified into:

1. **Finite Population**. Set of a limited number of elements, such as the number of species at Galapagos Islands, the number of students at the University of Guayaquil, the number of workers in an automobile factory [10–12]
2. **Infinite Population**. It has a large number of components, such as the set of species in the animal kingdom.

L. Cevallos-Torres and M. Botto-Tobar, *Problem-Based Learning: A Didactic Strategy in the Teaching of System Simulation*, Studies in Computational Intelligence 824, https://doi.org/10.1007/978-3-030-13393-1_2

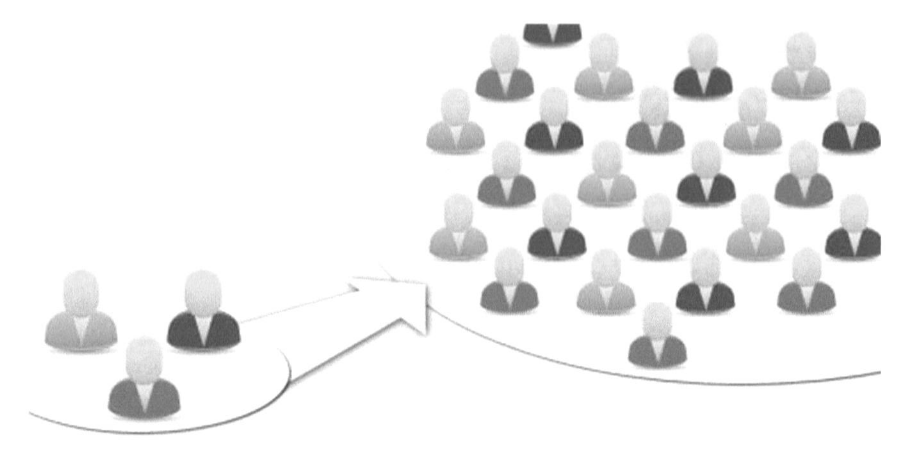

Fig. 2.1 Population [6]

3. **Real Population**. It is the whole group of concrete elements, such as people who are engaged in artistic activities in Europe.
4. **Hypothetical Population**. It is the set of possible imaginable situations in which an event may occur, such as the ways a person responds to a catastrophe as the earthquake in Ecuador in 2016 [13, 14].
5. **Stable Population**. Its features do not present variations, or if they do exists but in a small amount, they may be negligible, such as the earth rotation, the light speed [15].
6. **Unstable Population**. It contains values constantly change. This change occur in time or space.
7. **Random Population**. It presents changes in its heats due to chance, without an apparent cause, such as the variability found in the content of the product, whether its shape, size, length, among other characteristics [16].
8. **Dependent Population**. It changes its values due to a determined and measured cause. Its dependence can be *total*, such as the variations obtained in a mathematical function, or linear regression; or *partial* when the cause influences in the dependent variable in a lower proportion, for example: the increase in sales from higher advertising spending. This last influence is not proportional [17–19].
9. **Binomial Population**. It seeks the presence or absence of a characteristic, for example, the presence of a pest in the cultivation of a product, normally represented in such a way that it can be said that there is the presence of such a pest (p), or the absence of the plague (q), where p, q represent probabilities [20].
10. **Polynomial Population**. It has several characteristics that must be defined, measured or estimated, such as obedience, intelligence and postgraduate students age of graduate department at the University of Guayaquil [16, 21].

2.1.2 *Population Elements*

When a specific work is carried out, it is convenient to distinguish between the **theoretical population**: set of elements to which the results are to be extrapolated, and the **population studied**: set of accessible elements in our study [8, 22].

Sometimes, it is possible to study each element that constitutes to the population through a census [23, 24]. In other words, the study of all the elements that conform to the population [25]. The realization of a census is not always possible, for different reasons: (a) *Economy*: It is usually a costly problem in time, money if the population is large; (b) *Destructive test*; or (c) *infinite population*.

In sampling, the population is understood as the whole universe, and it is necessary that it be well defined so that it is known what elements constitute it at all times. If the numbering of elements is done on the accessible or studied population, and not on the theoretical population, the process is called a *sampling frame* or *space* [26–28].

2.2 Sampling

Sampling is the process of selecting a set of individuals from a population in order to study them and characterize the total population. It is the method used to choose a *sample* from the population, since in most cases it is not possible to study the entire population, and a sample can represent its individuals. This sample must be representative of all features of all elements [29–31].

In other words, it is a procedure or technique to know the population based on a sample drawn from it. It is also the study or procedure to determine characteristics of a population based on the information given by the sample in its data, and because it represents to the population, the sampling can be used for everything related to quality control, production, salaries, and wages, etc [8, 28].

2.2.1 *Terminology*

- **Population**. Set of individuals from whom we want to obtain information.
- **Sampling units**. Number of elements of population, not overlapping, that will be studied. Every member of the population will belong to one and only one sampling unit.
- **Units of analysis**. It is an object or individual from which the information must be obtained.
- **Sample frame**. It is a list of units or sampling elements.
- **Sample**. Set of units or elements of analysis taken from the frame.

2.2.2 Sampling Types

According to [28, 32] "samples can be chosen through various techniques or procedures" [33]. These techniques are classified into two broad groups:

- Probabilistic sampling or random sampling.
- Non-probabilistic sampling or non-random sampling.

2.2.2.1 Probabilistic Sampling

It is the process of selecting individuals in such a way that each subject has a positive and independent probability of being chosen to integrate the sample. In other words, it consists of choosing a sample of a random population [34, 35]. It is subdivided into:

- Simple random sampling.
- Systematic sampling.
- Stratified sampling.
- Sampling by conglomerates.

Simple Random Sampling.
It is the sampling technique in which all the population elements have the same probability of being part of the sample According to [36] "Simple random sampling is a method of selection of n units taken from N, in such a way that each of the samples has the same probability of being chosen", meaning that, all those who form part of the selection may be chosen. This method ensures that all individuals who compose the target population have the same opportunity to be included in the sample. It means that the probability of selecting a subject to study X is independent (see Fig. 2.2) [37].
There are two forms of simple random sampling:

- **Sampling with replacement**. An element can be selected more than once in the sample. For that, an element is extracted from the population, it is observed and is returned to the population so that in this way can be done infinite extractions from the population although it is finite [39–41].
- **Sampling without replacement**. The extracted elements are not returned to the population until all the elements of the population that conform to the sample are extracted [42, 43].

Advantages

- The procedure is efficient if the population is not large.
- It is relatively easy and cheap to find the units show them.

Disadvantages

- It requires the identification and cataloging of the population, which is sometimes very expensive.

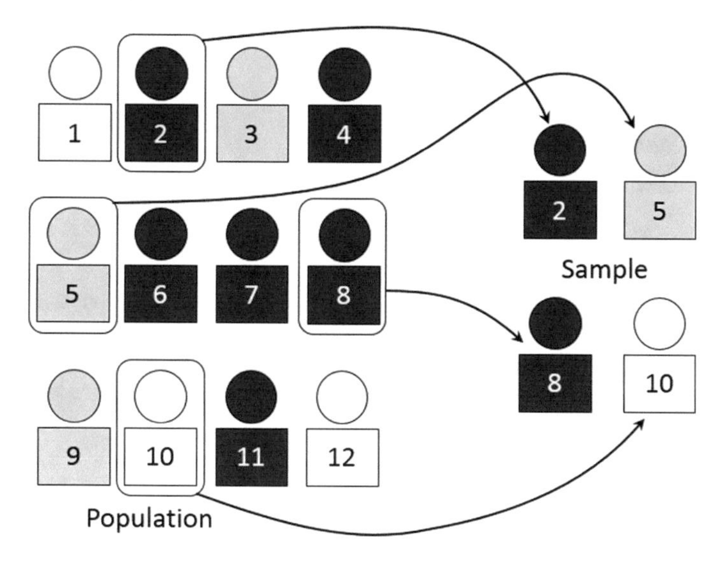

Fig. 2.2 Simple random sampling [38]

- It requires an equal selection probability for all the elements that make up the population.
- It requires a larger sample size than the other types of sampling.

When can I use this type of simple random sampling?

- It is recommended when the population is small.
- When there is a low level of heterogeneity in the data.
- When the population is located in a small space.
- When there is no previous information about the population.

Steps to get a simple random sampling. According to [28, 36] to make a simple random sample we must comply with the following steps:

1. Define the study population.
2. Enumerate all the analysis units that make up the population, assigning an identity number or identification.
3. Determine the optimal sample size for the study.
4. Select the sample systematically using a computer-generated random number table to ensure that you have a random order.

Simple random sampling formula:

$$n = \frac{N z_{\alpha/2}^2 pq}{(N-1)e^2 + z_{\alpha n}^2 pq} \tag{2.1}$$

where:

n = Sample size.
N = Population size.
$Z_{\alpha/2}$ = Confidence level.
$p * q$ = Proportional variance.
e = Error.
$p * q \quad e$ = Probability.

Exercise 1. Determine quality and service level. Suppose that you want to determine the quality and level of service offered by the Archivist Information Unit. So it is necessary to interview the different users that come to our file to know your opinion. How would calculate the sample size? For this case, it gets a frame master of 43,700 records corresponding to the visit log.

(a) Set the confidence level to 95% and an error of 5%.
$N = 43{,}700$
$P = 0.5$
$q = 0.5$
$Z_{\alpha/2} = 95\%$
$e = 0.05$
$Z_{\alpha/2} = 1.96$

$$n = \frac{(43{,}700)(1.96)^2(0.5)(0.5)}{(43{,}700 - 1)(0.5)^2 + (1.96)^2(0.5)(0.5)}$$
$$n = 38$$

(b) Set the confidence level to 90% and an error of 10%.
$N = 43{,}700$
$P = 0.5$
$q = 0.5$
$Z_{\alpha/2} = 90\%$
$e = 0.10$
$Z_{\alpha/2} = 1.65$

$$n = \frac{(43{,}700)(1.65)^2(0.5)(0.5)}{(43{,}700 - 1)(0.1)^2 + (1.65)^2(0.5)(0.5)}$$
$$n = 67.95$$

(c) Set the confidence level to 95% and an error of 10%.
$N = 43{,}700$
$P = 0.5$
$q = 0.5$
$Z_{\alpha/2} = 95\%$
$e = 0.10$
$Z_{\alpha/2} = 1.65$

$$n = \frac{(43{,}700)(1.96)^2(0.5)(0.5)}{(43{,}700-1)(0.1)^2+(1.96)^2(0.5)(0.5)}$$
$$n = 95.83$$

Exercise 2. Money draw. A specific company has generated significant profits for the sale of a certain product, the company's policy is to reward its employees with a voucher on its excellent performance, the company can not give that bonus to all its employees. Thus, an "X" amount of money is drawn between 4500. It is desired to know how many employees would choose to give them a voucher. What would be the optimal number of bonus winners?
Consider, for this calculation, a margin error of 20% and the confidence level of 96%.
$(1-\alpha) - 100\% = 96\%$
$1-\alpha = \frac{96\%}{100\%} \rightarrow 1-\alpha = 0.96 \rightarrow \alpha = 1-0.96$
$\alpha = 0.04 = 2.05$
$Z_{0.04/2} = Z_{0.02}$
$N = 4{,}500$
$P = 0.5$
$q = 0.5$
$Z_{\alpha/2} = 96\%$
$e = 0.20$
$Z_{\alpha/2} = 2.05$

$$n = \frac{(4{,}500)(2.05)^2(0.5)(0.5)}{(4{,}500-1)(0.2)^2+(2.05)^2(0.5)(0.5)}$$
$$n = 38.73$$

Exercise 3. Awards competition. The National Lottery of Guayaquil has launched a new sweepstakes, for which a contest is held. The shareholders of the National Lottery in conjunction with the planning department, elaborate 5,790 cartons. Therefore, it is desired to know from all the cartons that have been made, how many of them should be rewarded. To get this information, the specialist in the Statistics department suggests that a sample will determine the number of cartons. How many prize cartons should be awarded? Consider for this calculation, an error level of 6% and a confidence level of 90%.
$(1-\alpha) - 100\% = 90\%$
$1-\alpha = \frac{90\%}{100\%} \rightarrow 1-\alpha = 0.90 \rightarrow \alpha = 1-0.90$
$\alpha = 0.1 = 1.65$
$Z_{0.01/2} = Z_{0.05}$

$$n = \frac{(5{,}790)(1.65)^2(0.5)(0.5)}{(5{,}790-1)(0.06)^2+(1.65)^2(0.5)(0.5)}$$
$$n = 183$$

Exercise 4. Determining the consumption of a product. In a study carried out for the company "Great Milk", it was desired to determine in what proportion the children of a certain region in the Guayas province would take milk powder brand "Pediasure" at breakfast. It is known that there are 1,500 children, the company wants to know by a sample, the optimal amount of children who would try that product. To determine the sample size of children that consume Pediasure, consider a 10% error, with a confidence level of 95%.
$(1 - \alpha) - 100\% = 95\%$
$1 - \alpha = \frac{95\%}{100\%} \rightarrow 1 - \alpha = 0.95 \rightarrow \alpha = 1 - 0.95$
$\alpha = 0.5 = 1.96$
$Z_{0.05/2} = Z_{0.025}$

$$n = \frac{(1{,}500)(1.96)^2(0.5)(0.5)}{(1{,}500 - 1)(0.1)^2 + (1.96)^2(0.5)(0.5)}$$
$$n = 90$$

Exercise 5. Deterioration of a product. In a particular fruit shop located in the north of Guayaquil, it is bought daily 4000 oranges for sale. The fruit shop receives many complaints from its customers, because every time they acquire a certain amount of oranges, it turns out that some are damaged. The shop owner needs to know by using a sample the number of oranges that might be in bad condition.

Consider for this calculation, a 10% error level and a confidence level of 95%.
$(1 - \alpha) - 100\% = 99\%$
$1 - \alpha = \frac{99\%}{100\%} \rightarrow 1 - \alpha = 0.99 \rightarrow \alpha = 1 - 0.99$
$\alpha = 0.01 = 2.58$
$Z_{0.01/2} = Z_{0.005}$

$$n = \frac{(400)(2.58)^2(0.5)(0.5)}{(400 - 1)(0.1)^2 + (2.58)^2(0.5)(0.5)}$$
$$n = 117.73$$

Systematic random sampling.

It consists of choosing an initial individual randomly between the population and then selecting for the sample each nth individual available in the sample frame. Systematic sampling is a straightforward process, and it only requires the choice of a random individual. Hence, we consider it as trivial and fast, knowing that the results we obtain are representative of the population [43, 44].

Similar to simple random sampling: "the first individual is chosen at random, and the rest is conditioned by it, this method is straightforward to apply in practice and has the advantage that it does not need to have an elaborated survey framework" (see Fig. 2.3) [45, 46].

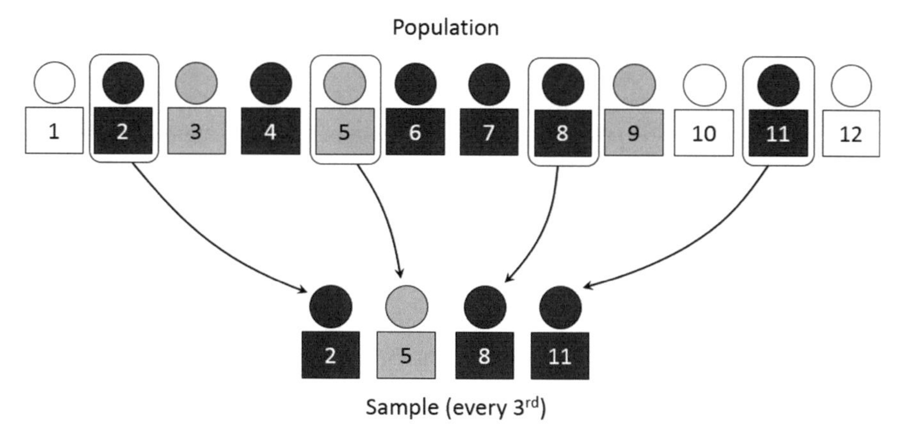

Fig. 2.3 Systematic sampling [47]

When can I use this type of simple random sampling?

- It is recommended when the population is large.
- When a list of the population elements can be available.
- When the elements of the population do not keep any periodicity with an important characteristic for the investigation.

Advantages

- Easy to apply.
- It is not always necessary to have a list of the whole population.
- When the population is ordered following a known trend, it ensures a coverage of units of all types.

Disadvantages

- If the sampling constant is associated with the phenomenon of interest, biased estimates can be found.

Steps. Concretely, the steps that we would continue to get a systematic sampling will be as follows:

1. Produce an ordered list with **n** individuals of the population, which would be the sampling frame.
2. Divide the sampling frame in n fragments, where n is the sample size desired. The fragments size will be

$$k = \frac{N}{n}$$

   where k gets the range or the lift coefficient.
3. **The start number**: It gets a random integer A, less than or equal to the interval. This number will be the first individual that we select for the sample within the first fragment from the divided population.

4. **The remaining n − 1 selection**: It selects the following individuals from the individual picked randomly through an arithmetic succession, by selecting individuals from the rest of the fragments into which we have divided the sample that they occupy the same position as the original subject. This amounts to saying that we will select individuals.

$$A, A + K, A + 2K, A + 3K, \ldots, A + (n - 1)K$$

Example. Suppose we have a sampling frame of 100 individuals, and it is desired to obtain a sample of 25. We divided into first place the sampling frame in 25 fragments of 4 individuals. Then, we select a random number between 1 and 4 to extract the first individual at random from the first fragment: for example 2. From this individual, the sample will be by removing individuals from the list with four units intervals.

$$k = \frac{N}{n}$$

$N = 100$
$n = 25$

$$k = \frac{N}{n} \rightarrow k = \frac{100}{25} \rightarrow k = 4$$

Exercise 1. Records of departments. It has a population of 3,000 records of some departments at the University of Guayaquil, and it is desired to take a systematic sample of 26 records.

$$k = \tfrac{3{,}000}{26} \rightarrow k = 115$$

Exercise 2. Survey. It is desired to know the opinion about a teacher of a lecture with 60 people. These people are sorted in alphabetical order according to lecture students list. To do the survey, 12 people are selected. Thus, $N = 60$, and $n = 12$. The fixed interval is 5.

$$k = \tfrac{60}{12} \rightarrow k = 5$$

Exercise 3. Dinner. From a total of 200 different dishes, it is desired to reduce to samples of 5 elements of dishes to show in a dinner.

$$k = \tfrac{200}{5} \rightarrow k = 40$$

Exercise 4. Shoe models. It is desired to divide a sample of 500 shoe types in groups of 50 elements to choose which n show in a shop showcase.

$$k = \tfrac{500}{50} \rightarrow k = 10$$

Exercise 5. Cell phone models. From a total of 400 different cell models, it is necessary to divide into samples of 20 to choose which to put in a promotion.

$$k = \frac{400}{20} \rightarrow k = 20$$

Exercise 6. Employers. A company has 120 employees, and it is desired to extract a sample of 30 of them.

$$k = \frac{120}{30} \rightarrow k = 4$$

Exercise 7. Teenagers. It is desired to extract a sample with an error of 10% and 95% of confidence from a population of 1500 adolescents.

$$n = \frac{(1{,}500)(1.96)^2(0.5)(0.5)}{(1{,}500-1)(0.1)^2+(1.96)^2(0.5)(0.5)} \rightarrow n = 91$$
$$k = \frac{1{,}500}{91} \rightarrow k = 16$$

Stratified sampling.

It consists of dividing the entire population under study into different subgroups or disjoint strata so that an individual can only belong to a stratum. In other words, "Stratified sampling is one in which we divide the population into subgroups or strata. Stratification can be based on a wide variety of attributes or characteristics found in the population as a profession, stature, age, gender, etc." [48–50].

"In stratified sampling, a partition of the population is first performed in subpopulations called strata, and sampling is carried out independently within each stratum" (see Fig. 2.4) [51, 52].

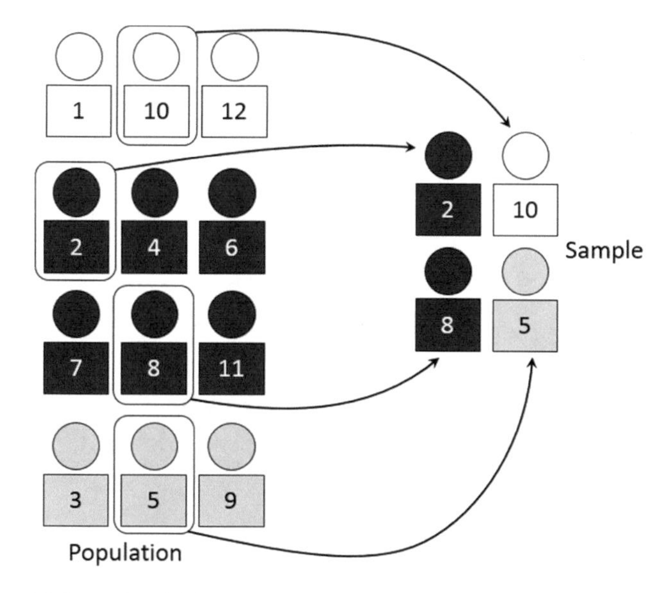

Fig. 2.4 Stratified sampling [53]

Table 2.1 Registration in Departments

Strata	Stratum 1	Stratum 2	Stratum 3	Stratum 4	Total population
Size	1,192	220	1,131	457	3,000

Table 2.2 Students

Strata	Stratum 1	Stratum 2	Stratum 3	Stratum 4	Total population
Size	950	230	315	505	2,000

Table 2.3 Products

Strata	Stratum 1	Stratum 2	Stratum 3	Stratum 4	Total population
Size	100	167	125	58	450

Exercise 1. Registration in Departments. It has a population of 3,000 records of some departments at the University of Guayaquil, and it is desired to take a systematic sample of 26 records (see Table 2.1).

$$n_i = n\frac{N_i}{N} \quad where \quad N = 3{,}000, \quad n = 26$$
$$n_1 = 26 * \frac{1192}{3{,}000} = 10.33 = 11$$
$$n_2 = 26 * \frac{220}{3{,}000} = 1.91 = 2$$
$$n_3 = 26 * \frac{1131}{3{,}000} = 9.81 = 9$$
$$n_4 = 26 * \frac{457}{3{,}000} = 3.96 = 4$$
$$\overline{26}$$

Exercise 2. Students sample. It has a population of 2,000 records, and it is desired to take a systematic sample of 100 records (see Table 2.2).

$$n_i = n\frac{N_i}{N} \quad where \quad N = 2{,}000, \quad n = 100$$
$$n_1 = 100 * \frac{950}{2{,}000} = 47.5 = 47$$
$$n_2 = 100 * \frac{230}{2{,}000} = 11.5 = 11$$
$$n_3 = 100 * \frac{315}{2{,}000} = 15.75 = 17$$
$$n_4 = 100 * \frac{505}{2{,}000} = 25.25 = 25$$
$$\overline{100}$$

Exercise 3. Products sample. It is desired to extract a sample that contains 450 elements. How many elements are to be allocated to each stratum? (see Table 2.3)

Table 2.4 Unemployed people

Strata	Stratum 1	Stratum 2	Stratum 3	Stratum 4	Total population
Size	45	95	39	691	870

$$
\begin{aligned}
n_i &= n\tfrac{N_i}{N} \quad where \quad N = 450, \quad n = 50\\
n_1 &= 50 * \tfrac{100}{450} = 11.11 = 11\\
n_2 &= 50 * \tfrac{167}{450} = 18.55 = 19\\
n_3 &= 50 * \tfrac{125}{450} = 13.88 = 14\\
n_4 &= 50 * \tfrac{58}{450} = 6.44 = 6\\
& \overline{50}
\end{aligned}
$$

Exercise 4. Unemployed people sample. It has a population of 870 unemployed people in Ecuador, and it is desired to take a stratified sample of 15 first records (see Table 2.4).

$$
\begin{aligned}
n_i &= n\tfrac{N_i}{N} \quad where \quad N = 870, \quad n = 15\\
n_1 &= 15 * \tfrac{45}{870} = 0.77 = 1\\
n_2 &= 15 * \tfrac{95}{870} = 1.63 = 1\\
n_3 &= 15 * \tfrac{39}{870} = 0.67 = 1\\
n_4 &= 15 * \tfrac{691}{870} = 11.91 = 12\\
& \overline{15}
\end{aligned}
$$

Exercise 5. Professional people sample. It is desired to extract a sample of people who are professionals, for instance: engineers, architects, lawyers, etc., from a population of 9345 elements. How many elements have to be assigned to each stratum, knowing that n is 893? (see Table 2.5).

$$
\begin{aligned}
n_i &= n\tfrac{N_i}{N} \quad where \quad N = 9,345, \quad n = 893\\
n_1 &= 893 * \tfrac{4,528}{9,345} = 432.69 = 433\\
n_2 &= 893 * \tfrac{1,333}{9,345} = 127.38 = 127
\end{aligned}
$$

Table 2.5 Professional people

Strata	Stratum 1	Stratum 2	Stratum 3	Stratum 4	Total population
Size	4,528	1,333	1,428	2,056	9,345

$$n_3 = 893 * \frac{1{,}428}{9{,}345} = 136.45 = 136$$
$$n_4 = 893 * \frac{2{,}056}{9{,}345} = 196.46 = 197$$
$$\overline{893}$$

Exercise 6. Cellphone models sample. It has a population of 500 models of cellphone in Guayaquil, and it is desired to take a stratified sample of 45 first models (see Table 2.6).

$$n_i = n\frac{N_i}{N} \quad where \quad N = 500, \quad n = 45$$
$$n_1 = 45 * \frac{45}{500} = 4.05 = 4$$
$$n_2 = 45 * \frac{85}{500} = 7.65 = 8$$
$$n_3 = 45 * \frac{220}{500} = 19.8 = 20$$
$$n_4 = 45 * \frac{150}{500} = 13.5 = 13$$
$$\overline{45}$$

Exercise 7. Population sample. It is desired to extract a global sample of 500 elements from a population divided into strata as shown in the Table 2.7. How many items are to be allocated to each stratum?

$$n_i = n\frac{N_i}{N} \quad where \quad N = 2{,}057, \quad n = 500$$
$$n_1 = 500 * \frac{466}{2{,}057} = 113.27 = 113$$
$$n_2 = 500 * \frac{125}{2{,}057} = 30.38 = 30$$
$$n_3 = 500 * \frac{549}{2{,}057} = 133.44 = 134$$
$$n_4 = 500 * \frac{917}{2{,}057} = 222.9 = 223$$
$$\overline{500}$$

Table 2.6 Cellphone models

Strata	Stratum 1	Stratum 2	Stratum 3	Stratum 4	Total population
Size	45	85	220	150	500

Table 2.7 Population

Strata	Stratum 1	Stratum 2	Stratum 3	Stratum 4	Total population
Size	466	125	549	917	2,057

Cluster Sampling.

It is a technique used when there are relatively homogeneous "natural" clusters in a statistical population. "It is a probabilistic sampling design by conglomerate; a conglomerate is considered a grouping of elements that present characteristics similar to the whole population" [34].

"It is a technique that takes advantage of the existence of groups or conglomerates in the population that correctly represent to the total population about the characteristic that is desired to measure where the groups contain all the population variability" [54]. "It consists of randomly choosing certain neighborhoods or conglomerates within a region, city, commune, etc., and then choosing smaller units such as blocks, streets, etc., and finally smaller ones, such as schools, clinics, homes" (see Fig. 2.5) [55, 56].

Advantages and Disadvantages

- The main advantage is operational type: select a conglomerate to study tends to be easier and more economical than a simple random or systematic.
- Interestingly, it is usual to make studies over the Internet that are still thinking in terms of study only a few geographical areas, when in reality via the Internet we do not get any operating profit; on the contrary, incur higher risk of having lower precision by differences among the regions studied and the rest of the population. This practice is an unjustified inheritance of techniques that were good in personal interviews, but who are not using other methodologies.
- The main drawback when using sampling cluster run a major risk: that the conglomerates are not homogeneous among them. For example, we have the case over

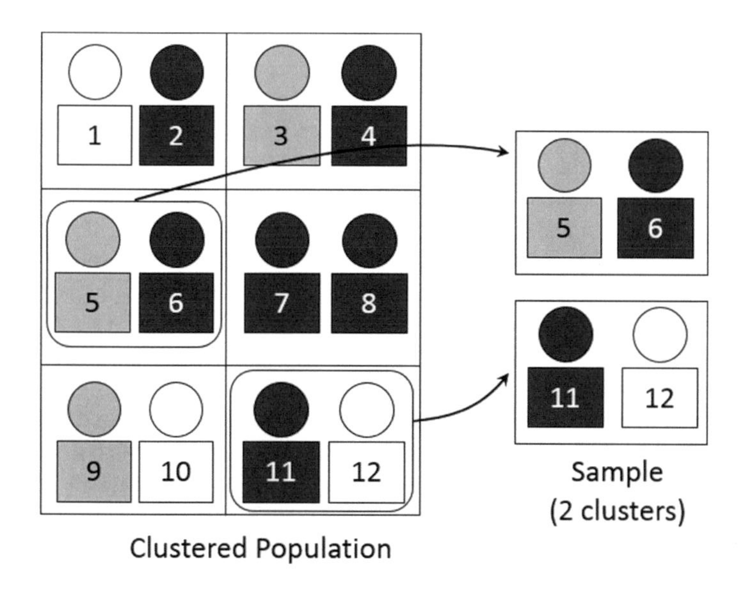

Fig. 2.5 Cluster sampling [57]

smoking in Ecuador, and it could happen in one of the provinces there is more likely to smoke, for being a more urban region, for cultural reasons, for commercial purposes.

Exercise. It is desired to know the degree of work satisfaction the professors' in high school, so that it is needed a sample of 700 subjects. Given the difficulty of accessing to these individuals, it is decided to make a sample by conglomerates. Knowing that the number of professors per institute is approximately 35. Therefore, the steps to follow would be the following:

- Collect a list of all institutes.
- Choose by using simple or systematic random sampling the 20 institutes (700/35 = 20) that will provide the 700 teachers it needs.

Proposed Exercises

1. A school has 120 high school students listed from 1 to 120, and it is desired to extract a sample of 30 students.

   (a) Through the simple random sampling.
   (b) Through the systematic random sampling.

2. To obtain a sample of students from a school to apply a survey. The enrolled students were listed and a total of 700 was obtained, and a sample of 75 was required.

   (a) Through the simple random sampling.
   (b) Through the systematic random sampling.

3. A student organization wants to estimate the proportion of students who are in favor of a school disposition. There is a list of the names and addresses of the 645 students enrolled in this quarter. It is desired to determine the first 10 students who will be selected using a simple random sampling.
4. It supposes it is going to take a simple random sample of 12 of 372 physicians in a given city. A medical organization gives the doctors names.
5. If it has to select a sample of 40 people from a community of 500 inhabitants to provide them with a survey of the health services they receive. The inhabitants are divided into five colonies: San Miguel, San Rafael, San Vicente, San Marcos, San Pedro where the size of each stratum is: 100, 150, 50, 125, 75, respectively. It is desired to determine the samples per stratum.
6. There is a list of pre-numbered checks with a population of 800, and it is desired to extract a systematic sample of 40.
7. To estimate the average of a certain variable, it has divided the data into four strata. Each of these strata contains the number of items indicated in Table 2.8: It is desired to extract a sample that globally contains 150 elements. How many items have to be assigned to be selected from each stratum?
8. Considering as a population the odd natural numbers from 1 to 40 estimate the population average using a random sample of 8 elements using the following sampling techniques:

Table 2.8 Exercise

Strata	Stratum 1	Stratum 2	Stratum 3	Stratum 4
Size	110	512	653	221

(a) The simple random sampling.
(b) The systematic random sampling.

9. In a factory of electronic articles usually, 10% of the articles present some manufacturing defect. It is desired to estimate the proportion of defects electronic articles from 2,000 items ready to be shipped. How many items should be chosen from the lot if it the confidence values is 95% and the estimation error is no greater than 0.05?

References

1. Traulsen, Arne, Jens Christian Claussen, and Christoph Hauert. 2005. Coevolutionary dynamics: from finite to infinite populations. *Physical Review Letters* 95 (23): 238701.
2. Paterakis, Michael, Leonidas Georgiadis, and Panayota Papantoni-Kazakos. 1987. On the relation between the finite and the infinite population models for a class of raa's. *IEEE Transactions on Communications* 35 (11): 1239–1240.
3. Shorrocks, Anthony F. 1984. Inequality decomposition by population subgroups. *Econometrica: Journal of the Econometric Society* 1369–1385.
4. Imbens, Guido W., and Donald B Rubin. 2015. *Causal inference in statistics, social, and biomedical sciences*. Cambridge: Cambridge University Press.
5. Pett, Marjorie A. 2015. *Nonparametric statistics for health care research: Statistics for small samples and unusual distributions*. New York: Sage Publications.
6. Kalla, Siddharth. 2009. Statistical significance, Sample Size and Expected Effects.
7. Camacho-Sandoval, Jorge. 2007. Investigación, poblaciones y muestra. *Acta Médica Costarricense* 49 (1): 11–12.
8. Ventura-León, José Luis. 2017. ¿población o muestra?: Una diferencia necesaria. *Revista Cubana de Salud Pública* 43 (4).
9. Herrera Llanos, Wilson. 2011. La población (segundo elemento constitutivo del estado colombiano). *Revista de Derecho* 19 (19).
10. Gallego, José A Mayor. 2011. Muestreo en poblaciones finitas.
11. Ojeda Mario Miguel. 2003. La inferencia en muestreo de poblaciones finitas y el analisis de datos de encuestas. Technical report, e-libro, Corp.
12. Belaire, Begoña Font. 1999. Una revisión de diferentes aportaciones al diseño en poblaciones finitas. *Qüestiió: quaderns d'estadística i investigació operativa*, 23 (1): 3–39.
13. Betanzos-Mendoza, Esteban. 1975. competencia entre plantas y la genetica de poblaciones.-i. estimacion de medias y varianzas en una poblacion hipotetica. *Agricultura tecnica en Mexico*.
14. Badii, M., and A. Guillen. 2009. Decisiones estadísticas: Bases teóricas:(statistical decision making: Theoretical basis). *International Journal of Good Conscience* 5: 185–207.
15. Coale, Ansley J., Paul Demeny, and Barbara Vaughan. 2013. *Regional Model Life Tables and Stable Populations: Studies in Population*. New York: Elsevier.
16. Schoen, Robert. 2013. *Modeling multigroup populations*. Berlin: Springer Science & Business Media.
17. Albarrán Lozano, Irene, and Pablo Alonso González. 2009. La población dependiente en españa: estimación del número y coste global asociado a su cuidado. *Estudios de economía* 36 (2): 127–163.

18. Abellán García, Antonio, Cecilia Esparza Catalán and Julio Pérez Díaz. 2011. Evolución y estructura de la población en situación de dependencia.
19. Escolano, Antonio Alegre, Mercedes Ayuso Gutiérrez, Montserrat Guillén Estany, Malena Monteverde Verdenelli, and Enrique Pociello García. 2005. Tasa de dependencia de la población española no institucionalizada y criterios de valoración de la severidad. *Revista Española de Salud Pública* 79: 351–363.
20. Escamilla, Juan Bacilio Guerrero, Laura Miriam Franco Sánchez, and Sonia Bass Zavala. 2018. Modelo probabilístico para predecir la dinámica de la tasa de delincuencia en michoacán. *Ciencia Ergo Sum*, 25 (1): 1–38.
21. Ávila-Curiel, A., T. Shamah, L. Barragán, A. Chávez, M.A. Avila, and L. Juárez. 2004. Índice epidemiológico de nutrición infantil basado en un modelo polinomial de los valores de puntuación z del peso para la edad. *Archivos Latinoamericanos de Nutricion* 54 (1): 50–57.
22. Quitian, Hoover, Rafael E Ruiz-Gaviria, Carlos Gómez-Restrepo, and Martin Rondón. 2016. Pobreza y trastornos mentales en la población colombiana, estudio nacional de salud mental 2015. *Revista Colombiana de Psiquiatría*, 45: 31–38.
23. Prieto, Mercedes. 2015. El estado ecuatoriano a mediados del s. xx: el censo, la población y la familia indígena. *European Review of Latin American and Caribbean Studies/Revista Europea de Estudios Latinoamericanos y del Caribe*, 29–46.
24. Mazurek, Hubert. 2016. El censo en bolivia, una herramienta para el desarrollo. *Tinkazos-Revista Boliviana de Ciencias Sociales* 15 (32).
25. Marmolejo Carlos Carlos, and Jorge Cerda Troncoso. 2017. El comportamiento espacio-temporal de la población como instrumento de análisis de la estructura urbana: el caso de la barcelona metropolitana. *Cuadernos Geográficos* 56 (2): 111–133.
26. Badii, M.H., J. Castillo, and A. Guillen. 2017. Tamaño óptimo de la muestra. *InnOvaciOnes de NegOciOs*, (09).
27. Orozco, Carlos Andrés Tobón, and José Rubiel Bedoya Sánchez. 2017. Influencia de la asimetría en el tamaño de la muestra para el cumplimiento del teorema central del límite en distribuciones continúas. *Scientia et technica* 22 (4): 398–402.
28. Otzen, Tamara, and Carlos Manterola. 2017. Técnicas de muestreo sobre una población a estudio. *International Journal of Morphology* 35 (1): 227–232.
29. Sabino, Carlos. 2014. *El proceso de investigación*. Editorial Episteme.
30. León, Luis Piña, Jorge D'Espaux Salgado, and Hugo de Rojas Gómez. 2018. Técnicas de muestreo aplicadas a la actividad empresarial: la auditoría (11). *Revista Economía y Desarrollo (Impresa)*, 148(2).
31. Mederos, Jose Almeida Cordero, Nexys Cabrera Padrón, Idanis Caraballo Castro, and Grisel Manso Silva. 2015. El muestreo estadístico, herramienta para proteger la objetividad e independencia de los auditores internos en las empresas cooperativas. *Cooperativismo y desarrollo*, 3 (1): 36–45.
32. Espinoza, I. 2016. *Tipos de muestreo*. Honduras: Universidad de Ciencias Medicas.
33. Mantilla Vargas, Farid A. 2015. Técnicas de muestreo: Un enfoque a la investigación de mercados.
34. Ochoa, C. 2015. Muestreo probabilístico o no probabilístico. *Blog la actualidad sobre la investigación por internet de Netquest. Recuperado de:* http://www.netquest.com/blog/es/muestreo-probabilistico-o-noprobabilistico-ii.
35. Carrasquedo, K. 2017. Muestreo probabilístico y no probabilístico. *Recuperado de https://www. gestiopolis. com/muestreo-probabilistico-y-no-probabilistico.*
36. Moreno, Amable. 2017. Dificultades en la comprensión del concepto de muestra aleatoria simple en estudiantes universitarios.
37. Al Ghayab, Hadi Ratham, Yan Li, Shahab Abdulla, Mohammed Diykh, and Xiangkui Wan. 2016. Classification of epileptic eeg signals based on simple random sampling and sequential feature selection. *Brain Informatics* 3 (2): 85–91.
38. Commons, Wikimedia. 2016. File:simple random sampling.png—wikimedia commons, the free media repository. [Accessed 11 November 2018].

39. Ben-Hamou, Anna, Yuval Peres, Justin Salez, et al. 2018. Weighted sampling without replacement. *Brazilian Journal of Probability and Statistics* 32 (3): 657–669.
40. Shabbir, Javid, Abdul Haq, and Sat Gupta. 2014. A new difference-cum-exponential type estimator of finite population mean in simple random sampling. *Revista Colombiana de Estadística* 37 (1): 199–211.
41. Rivest, Ronald L. 2018. Consistent sampling with replacement. arXiv:1808.10016.
42. Sengupta, S. 2016. On comparisons of with and without replacement sampling strategies for estimating finite population mean in randomized response surveys. *Sankhya B* 78 (1): 66–77.
43. Shamir, Ohad. 2016. Without-replacement sampling for stochastic gradient methods. In *Advances in neural information processing systems*, pp. 46–54.
44. Saby, Nicolas, D.J. Brus, Hakima Boukir, and Vera Laetitia Mulder. 2015. Approximating the sampling variance of means estimated from systematic random sample data of the french soil monitoring network. *Pedometrics 2015, Cordoue, ESP, 2015-09-14-2015-09-18.*
45. Bhagat, M., S. Qureshi, S. Kembhavi, G. Chinnaswamy, T. Vora, M. Prasad, L. Sidddhartha, N. Khanna, and M. Ramadwar. 2018. Prospective study of systematic retroperitoneal lymph node sampling for wilms tumors and comparison with random lymph node sampling. In *PEDIATRIC BLOOD & CANCER*, vol. 65, S466–S466. NJ USA: WILEY 111 RIVER ST, HOBOKEN 07030-5774.
46. Aune-Lundberg, Linda, and Geir-Harald Strand. 2014. Comparison of variance estimation methods for use with two-dimensional systematic sampling of land use/land cover data. *Environmental Modelling & Software* 61: 87–97.
47. Commons Wikimedia. 2016. File:systematic sampling.png — wikimedia commons, the free media repository. [Accessed 11 November 2018].
48. Bayne, Michael G., and Arindam Chakraborty. 2018. Development of composite control-variate stratified sampling approach for efficient stochastic calculation of molecular integrals. arXiv:1804.01197.
49. Koyuncu, Nursel, and Cem Kadilar. 2016. Calibration weighting in stratified random sampling. *Communications in Statistics-Simulation and Computation* 45 (7): 2267–2275.
50. Shields, Michael D., Kirubel Teferra, Adam Hapij, and Raymond P. Daddazio. 2015. Refined stratified sampling for efficient monte carlo based uncertainty quantification. *Reliability Engineering & System Safety*, 142: 310–325.
51. Tipton, Elizabeth, Larry Hedges, Michael Vaden-Kiernan, Geoffrey Borman, Kate Sullivan, and Sarah Caverly. 2014. Sample selection in randomized experiments: A new method using propensity score stratified sampling. *Journal of Research on Educational Effectiveness* 7 (1): 114–135.
52. Jing, Liping, Kuang Tian, and Joshua Z. Huang. 2015. Stratified feature sampling method for ensemble clustering of high dimensional data. *Pattern Recognition*, 48 (11): 3688–3702.
53. Commons, Wikimedia. 2016. File:stratified sampling.png—wikimedia commons, the free media repository. [Accessed 11 November 2018].
54. Mauricio Bustamante Jamid, and Sandra Valbuena Antolinez. 2015. Modelo experimental con bloques aleatorios simples y análisis multivariado para el mejoramiento de procesos orgánicos en la agroindustria. *Revista EAN* 78: 12–19.
55. Gómez Martínez, Freddy, Alina María Ruiz Piedra, Edilberto Gonzáles Ochoa, and Edelmira Belkis Soca Guevara. 2017. Selección de sintomáticos respiratorios en la habana de enero a abril del 2016 utilizando el muestreo por conglomerado. *Revista Cubana de Informática Médica*, 9 (2): 144–150.
56. Overgaard, Hans, Neal Alexander, Juan Felipe Jaramillo, Víctor Alberto Olano, Sandra Vargas, Diana Sarmiento, Audrey Lenhart, Razak Seidu, Thor Axel Stenström, and María Inés Matiz. 2015. Control de diarrea y dengue en escuelas primarias rurales de colombia: protocolo de estudio para un ensayo aleatorio y controlado por conglomerados. *Revista Salud Bosque*, 4 (1): 75–90.
57. Commons, Wikimedia. 2016. File:cluster sampling.png—wikimedia commons, the free media repository. [Accessed 11 November 2018].

Chapter 3
Pseudo-Random Numbers and Congruential Methods

The pseudo numbers are the essential basis of the simulation. Usually, all randomness involved in the model is obtained from a random number generator that produces a succession of values that are supposed to be realizations of a sequence of independent random variables and identically distributed uniforms U (0, 1). To be more explicit about the use of pseudo numbers, we will analyze concepts such as mixed or linear congruence method, multiplicative congruence method, additive congruence method.

3.1 Pseudo-Random Numbers

Real systems often have time and quantity values that vary within a range and according to a specific density function, defined by a probability distribution. For example, if the time that takes a machine to process a part is distributed between 2.2 and 4.5 min, this will be defined as a probability distribution in the simulation model [1–3].

In the simulation experiments, it is necessary to generate values for the random variables represented these using probability distributions. In order to generate stochastic inputs (probabilistic) for a simulation model, you must have a pseudo number generator [1, 4–6].

Pseudo numbers are numbers generated in a process that seems to produce random numbers. A pseudo-random number is only the value of a random variable x that has a uniform probability distribution defined in the range (0, 1) [7–11].

L. Cevallos-Torres and M. Botto-Tobar, *Problem-Based Learning: A Didactic Strategy in the Teaching of System Simulation*, Studies in Computational Intelligence 824, https://doi.org/10.1007/978-3-030-13393-1_3

3.2 Pseudo Number Properties

It is desirable that uniform pseudo numbers possess the following characteristics [12, 13]:

1. Uniformly distributed.
2. Statistically independent.
3. Reproducible.
4. Long period.
5. Generated by a quick method.
6. Generated by a method that does not require a lot of storage capacity of the computer.

A random (pseudo) number generator is a structure $G = (X; x0; T; U; g)$, where X is a finite set of states, $x0 \in X$ is the initial state (seed), the application $T : X- > X$ is the transition function, U is the finite set of possible observations, and $G : X- > U$ is the output function [8, 14].

Example.

To get a pseudo-random number in Excel, we must use the $rand()$ function. It has no arguments, so it will be enough to place its name followed by both parentheses. It has entered the function in A1 cell, and as a result, a random number will be obtained (see Fig. 3.1) [11, 15–18].

If it is desired to generate more random numbers; it would be enough to copy the formula to other cells (See Figure). As it can be noticed in Fig. 3.2, the value of A1 cell has changed after the formula has copied down. This is because the random function is re-calculated every time that there is a change in the sheet, and therefore, it will have a new value in cell A1. Conversely, if it is desired to leave "fixed" the random numbers generated, it will be needed to copy them to another cell range by using the *Values* in *Paste* option. In this way, the random function in the new cells will be eliminated, and it will have the random numbers previously generated.

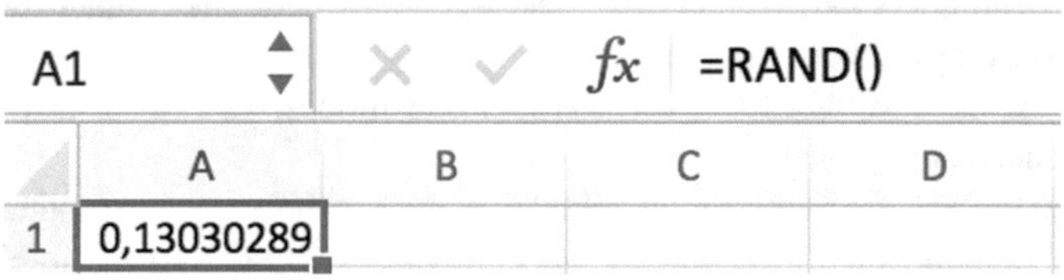

Fig. 3.1 Simulation of 1 pseudo-random number in Excel

A1 fx =RAND()

	A	B	C	D	E
1	0,57060679				
2	0,07998528				
3	0,29477297				
4	0,13076229				
5	0,41809184				
6	0,06478549				
7	0,10042331				
8	0,43130439				
9	0,75760118				
10	0,0782281				
11					
12					
13					

Fig. 3.2 Simulation of 10 pseudo-random numbers in Excel

3.3 Methods for Generating Pseudo Numbers

The independent variables in the mathematical model for a simulation are treated with random numbers (because they represent the variables that cannot be controlled). The "random" numbers through a Personal Computer (PC) are pseudo numbers generated by algorithms, and it is done from the following methods:

3.3.1 Manual Methods

They are the most straightforward and slowest methods. For examples, coin releases, dice, cards and roulettes. The numbers produced by these methods meet the statistical conditions mentioned earlier, though it is impossible to reproduce a sequence generated by these methods.

3.3.2 Random Number Tables

These numbers can be generated by employing a spreadsheet or by any generator of any programming language reason why their behavior is deterministic.

3.3.3 *Employing the Computer*

There are three methods to produce random numbers:

- External provision.
- Internal generation through a random physical process.
- Generation through a recurrence rule.

3.4 Arithmetic Methods to Generate Pseudo Numbers

3.4.1 *Mean Square Method*

The procedure for obtaining pseudo numbers with this type of generator is as follows:

- A seed is defined.
- The seed is raised squared.
- Depending on the number of digits is desired to have the pseudo-random number, It is taken from the central part of the resulting number in the previous step the number of digits required. If it is not possible to determine the central part, the number is completed by adding zeros at the beginning or end.
- It should be noted that pseudo numbers are desired between 0 and 1. Consequently, the result should be normalized, that is, if the numbers are two digits is normalized by dividing by 100, if it is three digits per thousand and so on.

Example. Generate 3 random numbers of 4 digits from a medium square generator using the seed number 445.

As it is desired 4-digit pseudo-numbers R_i, it will take the four digits from the central part of the seed square, as follows:

$$(445)^2 = 198025 = 9802 \quad \text{then} \quad R_1 = \frac{9802}{10000} = 0.9802$$

$$(9802)^2 = 96079204 = 0792 \quad \text{then} \quad R_2 = \frac{0792}{10000} = 0.0792$$

$$(792)^2 = 627264 = 2726 \quad \text{then} \quad R_3 = \frac{2726}{10000} = 0.2726$$

Note: As the pseudo numbers must be between 0 and 1, and are 4 digits, it normalizes by dividing between 10000.

3.4.2 Medium Product Method

This method is somewhat similar to the previous one, but it should start with two seeds each with k digits, the resulting number is taken as the central figures of the product of the two previous numbers. For example, taking as seeds to $X0 = 13$ and $X1 = 15$. The method would be as follows:

$$X_2 = (13 * 15) = 0195 = 19 \quad \text{then} \quad R_2 = \frac{19}{100} = 0.19$$

$$X_3 = (15 * 19) = 0285 = 28 \quad \text{then} \quad R_3 = \frac{28}{100} = 0.28$$

$$X_4 = (19 * 28) = 0532 = 53 \quad \text{then} \quad R_4 = \frac{53}{100} = 0.53$$

3.4.3 Modified Medium Product Method

This method consists of using a multiplication constant instead of a variable, that is $X_{n+1} = (K * X_n)$. It should be noted that the previous methods have relatively short periods, which are significantly affected by the initial values chosen, and are statistically unsatisfactory. It should also be noted that a generator with a short period is not used to make a considered number of simulation tests.

3.5 Congruence Methods

We have developed three methods of congruence to generate pseudo numbers, which are derived from the use of different versions of the fundamental relationship of congruence. The objective of each of the methods is the generation in a minimum time, of random number successions with maximum periods. The congruence methods are the mixed, the multiplicative and the additive [19, 20].

3.5.1 Mixed or Linear Congruence Method

A mixed or linear congruence method is an algorithm that allows obtaining a sequence of pseudo numbers calculated with a linear function defined as discontinuous pieces. Linear congruence generators generate a sequence of pseudo numbers in which the next pseudo-random number is determined from the last generated number, i.e., The pseudo-random number X_{n+1} is derived from the pseudo-random number X_n.

The recurrence ratio for the mixed congruence generator is $X_{n+1} = (aX_n + c)$ mod m, where:
x_0 is the seed $x_0 > 0$.
a the multiplier $(a > 0)$.
c is the additive constant $(c < 0)$.
m is the module $(m > x_0, m > a\, and\, m > c)$.
r_i is the random number.
$X_0, a, c > 0$.

This recurrence ratio tells us that X_{n+1} is the remainder of dividing to X_{n+c} between the module. The above means that the possible values of X_{n+1} are 0, 1, 2, 3... $m - 1$, that is, it represents the possible number of different values that can be generated.
$x_{i+1} = (ax_i + c)$ mod (m)
$x_i = x_0$
$a = 1 + 4k$
$m = 2^g$
$k = integer$
$g = integer$

$$r_i = \frac{x_i}{(m-1)}$$

Exercise 1. Mixed or linear congruence method.
Find a sequence of pseudo-random numbers for the following data:

x	k	a	c	g	m
6	8	33	5	2	4

$$a = 1 + 4k \qquad m = 2^G$$
$$a = 1 + 4(8) \qquad m = 2^2$$
$$a = 33 \qquad m = 4$$
$$x_i = (ax_i + c) \text{ mod (m)}$$

$x_1 = (33(6) + 5)$ mod (4)
$x_1 = 203$ mod (4)
$0.75 * 4 = 3$
$x_1 = 3$

$$r_1 = \frac{3}{4-1}$$
$$r_1 = \frac{3}{3}$$
$$r_1 = 1$$

$x_2 = (33(3) + 5) \bmod (4)$
$x_2 = 104 \bmod (4)$
$x_2 = 26.00$
$00.00 * 4 = 0$
$x_2 = 0$

$$r_2 = \frac{0}{4-1}$$

$$r_2 = \frac{0}{3}$$

$$r_2 = 0$$

$x_3 = (33(0) + 5) \bmod (4)$
$x_3 = 5 \bmod (4)$
$x_3 = 1$

$$r_3 = \frac{1}{4-1}$$

$$r_3 = \frac{1}{3}$$

$$r_3 = 0.33$$

$x_4 = (33(1) + 5) \bmod (4)$
$x_4 = 38 \bmod (4)$
$x_4 = 9.5$
$0.50 * 4 = 2$
$x_4 = 2$

$$r_4 = \frac{2}{4-1}$$

$$r_4 = \frac{2}{3}$$

$$r_4 = 0.66$$

$x_5 = (33(2) + 5) \bmod (4)$
$x_5 = 71 \bmod (4)$
$x_5 = 17.75$
$0.75 * 4 = 3$
$x_5 = 3$

$$r_5 = \frac{3}{4-1}$$

$$r_5 = \frac{3}{3}$$

$$r_5 = 1$$

The method finishes when it is observed that the numbers are repeated, in this case, if we look at the value of x_5. We can realize that it is equal to x_1, but what if I do to x_6, then we can see that it repeats, and it will be equal to x_2, as it is shown below:
$x_6 = (33(3) + 5) \bmod (4)$
$x_6 = 104 \bmod (4)$
$x_6 = 26.00$
$0.00 * 4 = 0$
$x_6 = 0$

$$r_6 = \frac{0}{4-1}$$

$$r_6 = \frac{0}{3}$$

$$r_6 = 0$$

x_1	3	0	1	2
r_i	1	0	0.33	0.66

When it is desired to build a random number generator to simulate the values of a random variable; it must choose the parameters in such a way that a long period is guaranteed so that all the simulation tests can be done. Therefore, it must take into account the following conditions:

- a must be an impair number, not divisible by 3 or by 5.
- c can be usually any constant; however, to ensure good results, it must be selected a so that, $a \bmod 8 = 5$ for a binary computer, or mod $200 = 21$ for the decimal computer.
- m must be the largest integer the computer. accepts.

According to Hull and Dobell, the best results for a mixed congruence generator on a binary computer are:
$c = 8 * A3$
$a =$ any integer
$X_0 =$ any impair integer
$m = 2b$ where $b > 2$, and m is accepted by the computer.

Exercise 2. Generate mixed pseudo numbers.
"NoviCompu" wants to make an inventory of its products, to choose the products the company needs to generate 17 random numbers. The specialist in the statistics department suggests to apply the linear congruence method, given the sequence of integers where: $x = 34$, $k = 15$, $c = 10$ and $G = 5$

x	k	a	c	g	m
34	15	61	10	5	32

$x_1 = (61(34) + 10) \bmod (32)$
$x_1 = 2084 \bmod (32)$
$x_1 = 65.125$
$x_1 = 0.125 * 32$
$x_1 = 4$

$$r_1 = \frac{4}{32 - 1}$$

$$r_1 = \frac{4}{31}$$

$$r_1 = 0.129$$

$x_2 = (61(4) + 10) \bmod (32)$
$x_2 = 254 \bmod (32)$
$x_2 = 7.9375$
$x_2 = 0.9375 * 32$
$x_2 = 30$

$$r_2 = \frac{30}{32 - 1}$$

$$r_2 = \frac{30}{31}$$

$$r_2 = 0.967$$

$x_3 = (61(30) + 10) \bmod (32)$
$x_3 = 1840 \bmod (32)$
$x_3 = 57.5$
$x_3 = 0.5 * 32$
$x_3 = 16$

$$r_3 = \frac{16}{32 - 1}$$

$$r_3 = \frac{16}{31}$$

$$r_3 = 0.516$$

$x_4 = (61(16) + 10) \bmod (32)$
$x_4 = 986 \bmod (32)$
$x_4 = 30.81245$
$x_4 = 0.81245 * 32$
$x_4 = 26$

$$r_4 = \frac{26}{32 - 1}$$

$$r_4 = \frac{26}{31}$$

$$r_4 = 0.839$$

$x_5 = (61(26) + 10) \bmod (32)$
$x_5 = 1596 \bmod (32)$
$x_5 = 49.875$
$x_5 = 0.875 * 32$
$x_5 = 28$

$$r_5 = \frac{28}{32 - 1}$$

$$r_5 = \frac{28}{31}$$

$$r_5 = 0.903$$

$x_6 = (61(28) + 10) \bmod (32)$
$x_6 = 1718 \bmod (32)$
$x_6 = 53.6875$
$x_6 = 0.6875 * 32$
$x_6 = 22$

$$r_6 = \frac{22}{32 - 1}$$

$$r_6 = \frac{22}{31}$$

$$r_6 = 0.709$$

$x_7 = (61(22) + 10) \bmod (32)$
$x_7 = 1352 \bmod (32)$
$x_7 = 42.25$
$x_7 = 0.25 * 32$
$x_7 = 8$

$$r_7 = \frac{8}{32-1}$$

$$r_7 = \frac{8}{31}$$

$$r_7 = 0.258$$

$x_8 = (61(8) + 10) \bmod (32)$
$x_8 = 498 \bmod (32)$
$x_8 = 15.5625$
$x_8 = 0.5625 * 32$
$x_8 = 18$

$$r_8 = \frac{18}{32-1}$$

$$r_8 = \frac{18}{31}$$

$$r_8 = 0.581$$

$x_9 = (61(18) + 10) \bmod (32)$
$x_9 = 1108 \bmod (32)$
$x_9 = 34.625$
$x_9 = 0.625 * 32$
$x_9 = 20$

$$r_9 = \frac{20}{32-1}$$

$$r_9 = \frac{20}{31}$$

$$r_9 = 0.645$$

$x_{10} = (61(20) + 10) \bmod (32)$
$x_{10} = 1230 \bmod (32)$
$x_{10} = 38.4375$

$x_{10} = 0.4375 * 32$
$x_{10} = 14$

$$r_{10} = \frac{14}{32 - 1}$$

$$r_{10} = \frac{14}{31}$$

$$r_{10} = 0.452$$

$x_{11} = (61(14) + 10) \bmod (32)$
$x_{11} = 864 \bmod (32)$
$x_{11} = 27.00$
$x_{11} = 0.00 * 32$
$x_{11} = 0$

$$r_{11} = \frac{0}{32 - 1}$$

$$r_{11} = \frac{0}{31}$$

$$r_{11} = 0$$

$x_{12} = (61(0) + 10) \bmod (32)$
$x_{12} = 10 \bmod (32)$
$x_{12} = 0.3125$
$x_{12} = 0.3125 * 32$
$x_{12} = 10$

$$r_{12} = \frac{10}{32 - 1}$$

$$r_{12} = \frac{10}{31}$$

$$r_{12} = 0.323$$

$x_{13} = (61(10) + 10) \bmod (32)$
$x_{13} = 620 \bmod (32)$
$x_{13} = 19.375$
$x_{13} = 0.375 * 32$
$x_{13} = 12$

$$r_{13} = \frac{12}{32 - 1}$$

$$r_{13} = \frac{12}{31}$$

$$r_{13} = 0.387$$

$x_{14} = (61(12) + 10) \bmod (32)$
$x_{14} = 742 \bmod (32)$
$x_{14} = 21.1875$
$x_{14} = 0.1875 * 32$
$x_{14} = 6$

$$r_{14} = \frac{6}{32 - 1}$$

$$r_{14} = \frac{6}{31}$$

$$r_{14} = 0.194$$

$x_{15} = (61(6) + 10) \bmod (32)$
$x_{15} = 376 \bmod (32)$
$x_{15} = 11.75$
$x_{15} = 0.75 * 32$
$x_{15} = 24$

$$r_{15} = \frac{24}{32 - 1}$$

$$r_{15} = \frac{24}{31}$$

$$r_{15} = 0.774$$

$x_{16} = (61(24) + 10) \bmod (32)$
$x_{16} = 1474 \bmod (32)$
$x_{16} = 40.0625$
$x_{16} = 0.0625 * 32$
$x_{16} = 2$

$$r_{16} = \frac{2}{32 - 1}$$

$$r_{16} = \frac{2}{31}$$

$$r_{16} = 0.064$$

$x_{17} = (61(2) + 10) \bmod (32)$
$x_{17} = 132 \bmod (32)$
$x_{17} = 4.125$
$x_{17} = 0.125 * 32$
$x_{17} = 4$

$$r_{17} = \frac{4}{32 - 1}$$

$$r_{17} = \frac{4}{31}$$

$$r_{17} = 0.129$$

Exercise 3. Generate mixed pseudo numbers.
"Acromax" needs to know if the product "XYZ' is expired. There is a stock of 150.000 elements, so it proceeds to take a sample, to perform this work has been considered, generate 257 random numbers, but only be shown Manually 11 numbers applying the linear congruence method. The following values give the sequence of numbers $X_{i+1} = (aX_i + c) \bmod (m)$

$$r_i = \frac{X_i}{m - 1}$$

x	a	c	g	m
89	385	78	9	512

The random numbers generated by the mixed congruence method are shown in Table 3.1.

Table 3.1 Random number generation by mixed congruence method

n	X_n	$(a\ X_{n+c}) \bmod m$	X_{n+1} (Remainder)	Random numbers (R_i)
1	89	34343 mod 512	39	0.07632
2	39	15093 mod 512	245	0.47945
3	245	94325 mod 512	117	0.22896
4	117	45123 mod 512	67	0.13111
5	67	25873 mod 512	273	0.53424
6	273	105105 mod 512	145	0.28375
7	145	55903 mod 512	0	0
8	0	78 mod 512	78	0.15264
9	78	30108 mod 512	412	0.80626
10	412	158698 mod 512	490	0.958904
11	490	188728 mod 512	312	0.61056

Exercise 4. Generate mixed pseudo numbers.
Movistar needs to recycle a large number of old generation cell phones, as some of its components are required as extra material, the company, want to take a sample, for which will generate random numbers, all this process is It will do so by applying the linear congruence method.

$$X_{i+1} = (aX_i + c) \bmod (m)$$

$$r_i = \frac{X_i}{m-1}$$

x	a	c	m
45	47	58	288

$x_1 = (47(45) + 58) \bmod (288)$ $r_1 = 0.5470383275$
$x_2 = (47(157) + 58) \bmod (288)$ $r_2 = 0.8257839721$
$x_3 = (47(237) + 58) \bmod (288)$ $r_3 = 0.881533101$
$x_4 = (47(253) + 58) \bmod (288)$ $r_4 = 0.4912891986$
$x_5 = (47(141) + 58) \bmod (288)$ $r_5 = 0.212543554$
$x_6 = (47(61) + 58) \bmod (288)$ $r_6 = 0.1567944251$
$x_7 = (47(45) + 58) \bmod (288)$ $r_7 = 0.5470383275$
As a result we have:

X_i	157	237	253	141	61	45	157
R_i	0.547	0.82578	0.882	0.491	0.2125	0.1568	0.547

where i = 1, 2, 3, 4, 5, 6, 7.

Exercise 5. Generate mixed pseudo numbers.
Find the sequence of pseudo-random numbers with the following data.

$$X_{i+1} = (aX_i + c) \bmod (m)$$

$$r_i = \frac{X_i}{m-1}$$

x	a	c	G	K	m
6	36	5	2 ·	5	4

$x_1 = (33 * 6 + 5) \bmod (4) = 3$ $r_1 = \frac{3}{3} = 1$
$x_2 = (33 * 3 + 5) \bmod (4) = 0$ $r_2 = \frac{0}{3} = 0$

$x_3 = (33 * 0 + 5) \bmod (4) = 1 \qquad r_3 = \frac{1}{3}$
$x_4 = (33 * 1 + 5) \bmod (4) = 2 \qquad r_4 = \frac{2}{3}$
$x_5 = (33 * 2 + 5) \bmod (4) = 3 \qquad r_5 = \frac{3}{3} = 1$

Exercise 6. Generate mixed pseudo numbers.
Find pseudo-random number sequence with the following data.

$$X_{i+1} = (aX_i + c) \bmod (m)$$

$$r_i = \frac{X_i}{m - 1}$$

x	a	c	G	K	m
3	2	9	3	5	8

$x_1 = (21 * 3 + 9) \bmod (8) = 0 \qquad r_1 = \frac{0}{7} = 0$
$x_2 = (21 * 0 + 9) \bmod (8) = 1 \qquad r_2 = \frac{1}{7} = 0.1428$
$x_3 = (21 * 1 + 9) \bmod (8) = 6 \qquad r_3 = \frac{6}{7} = 0.8571$
$x_4 = (21 * 6 + 9) \bmod (8) = 7 \qquad r_4 = \frac{7}{7} = 1$
$x_5 = (21 * 7 + 9) \bmod (8) = 4 \qquad r_5 = \frac{4}{7} = 0.5714$
$x_6 = (21 * 5 + 9) \bmod (8) = 2 \qquad r_6 = \frac{2}{7} = 0.2857$
$x_7 = (21 * 2 + 9) \bmod (8) = 3 \qquad r_7 = \frac{3}{7} = 0.4285$

3.5.2 *Multiplicative Congruence Method*

Computes a X_n sequence of non-negative integers, each of which is less than M by the relationship $X_{n+1} = aX_n(\bmod M)$. It is a special case of the congruence relationship in which $c = 0$, this method behaves satisfactorily statistically, i.e., the numbers generated by this method are uniformly distributed and are not correlated. This method has a maximum period of less than M, but conditions can be imposed on a and x_0, so that the maximum period is obtained. From a computational point of view it is the fastest of all [21–23].

Therefore, the recursive equation is:
$x_{i+1} = (ax_i) \bmod (m)$
$x_i = x_0$
$a = 1 + 4k$
$m = 2^g$
k = an integer number.
g = an integer number.

$$R_i = \frac{X_i}{m - 1}$$

where:
$x_0 > 0$ represents the seed and is a value selected by the researcher;
$a > 0$ is the multiplier;
m is the module, being $m > x_0, m > a$.

Exercise 1. Generate numbers pseudo congruence multiplicative.
Use the multiplicative congruence method to generate five possible random numbers.

$$x_{i+1} = (ax_i) \bmod (m)$$

$$R_i = \frac{X_i}{m - 1}$$

x	a	m
85	32	95

$x_1 = (32(85)) \bmod (95)$
$x_1 = 270 \bmod (95)$
$x_1 = 28.63157895$
$x_1 = 0.63157895 * 95$
$x_1 = 60$

$$r_1 = \frac{60}{95 - 1}$$

$$r_1 = \frac{60}{94}$$

$$r_1 = 0.6382978723$$

$x_2 = (32(6)) \bmod (95)$
$x_2 = 1920 \bmod (95)$
$x_2 = 20.21052632$
$x_2 = 0.21052632 * 95$
$x_2 = 20$

$$r_2 = \frac{20}{95 - 1}$$

$$r_2 = \frac{20}{94}$$

$$r_2 = 0.2127659574$$

$x_3 = (32(20)) \bmod (95)$
$x_3 = 640 \bmod (95)$
$x_3 = 6.736842105$
$x_3 = 0.736842105 * 95$
$x_3 = 70$

$$r_3 = \frac{70}{95 - 1}$$

$$r_3 = \frac{70}{94}$$

$$r_3 = 0.7446808511$$

$x_4 = (32(70)) \bmod (95)$
$x_4 = 2240 \bmod (95)$
$x_4 = 23.57894737$
$x_4 = 0.57894737 * 95$
$x_4 = 55$

$$r_4 = \frac{55}{95 - 1}$$

$$r_4 = \frac{55}{94}$$

$$r_4 = 0.585106383$$

$x_5 = (32(55)) \bmod (95)$
$x_5 = 1760 \bmod (95)$
$x_5 = 1.71587$
$x_5 = 0.71587 * 95$
$x_5 = 50$

$$r_5 = \frac{50}{95 - 1}$$

$$r_5 = \frac{50}{94}$$

$$r_5 = 0.5319148936$$

Exercise 2. Generate numbers pseudo congruence.
Find the sequence of pseudo-random numbers with the following data.

$$x_{i+1} = (ax_i) \bmod (m)$$

$$R_i = \frac{X_i}{m-1}$$

x	k	a	g	m
17	2	21	5	32

$a = 5 + 8k \quad m = 2^g$
$a = 5 + 8(2) \quad m = 2^5$
$a = 5 + 16 \quad m = 32$
$a = 21$

$x_1 = (21 * 17) \bmod (32) = 5 \quad r_1 = \frac{5}{31} = 0.1616$
$x_2 = (21 * 5) \bmod (32) = 9 \quad r_2 = \frac{9}{31} = 0.2903$
$x_3 = (21 * 9) \bmod (32) = 29 \quad r_3 = \frac{29}{31} = 0.9355$
$x_4 = (21 * 29) \bmod (32) = 1 \quad r_4 = \frac{1}{31} = 0.0322$
$x_5 = (21 * 1) \bmod (32) = 21 \quad r_5 = \frac{21}{31} = 0.6774$
$x_6 = (21 * 21) \bmod (32) = 25 \quad r_6 = \frac{25}{31} = 0.8064$
$x_7 = (21 * 25) \bmod (32) = 13 \quad r_7 = \frac{13}{31} = 0.4194$
$x_8 = (21 * 13) \bmod (32) = 17 \quad r_8 = \frac{17}{31} = 0.5484$

Exercise 3. Generate numbers pseudo congruence multiplicative.
Find the sequence of pseudo-random numbers with the following data.

$$x_{i+1} = (ax_i) \bmod (m)$$

$$R_i = \frac{X_i}{m-1}$$

x	k	a	g	m
13	3	29	4	16

$x_1 = (21 * 13) \bmod (16) = 9 \quad r_1 = \frac{9}{15} = 0.6$
$x_2 = (21 * 9) \bmod (16) = 5 \quad r_1 = \frac{5}{15} = 0.33$
$x_3 = (21 * 5) \bmod (16) = 1 \quad r_1 = \frac{1}{15} = 0.067$
$x_4 = (21 * 1) \bmod (16) = 13 \quad r_1 = \frac{13}{15} = 0.867$

Exercise 4. Generate numbers pseudo congruence multiplicative.
Perform the following exercise, by the multiplicative congruence method, to generate random numbers using the following data.

$$x_{i+1} = (ax_i) \bmod (m)$$

$$R_i = \frac{X_i}{m-1}$$

x	k	a	g	m
235	87	701	6	64

$a = 5 + 8k \quad m = 2^g$
$a = 5 + 8(87) \quad m = 2^6$
$a = 5 + 696 \quad m = 64$
$a = 701$

Exercise 5. Generate numbers pseudo congruence multiplicative.
Solve the following exercise, by the multiplicative congruence method to generate possible random numbers.

$$x_{i+1} = (ax_i) \bmod (m)$$

$$R_i = \frac{X_i}{m-1}$$

x	a	m
90	67	520

Exercise 6. Generate numbers pseudo congruence multiplicative.
Apply the multiplicative congruence method to generate random numbers using the following data:

$$x_{i+1} = (ax_i) \bmod (m)$$

$$R_i = \frac{X_i}{m-1}$$

x	k	a	g	m
432	123	989	10	1024

3.5.3 Additive Congruence Method

Calculates a succession of pseudo numbers using the ratio $x_{i+1} = X_n + X_{n-k} \mod (m)$. To use this method, you need K initial values, being K integer. The statistical properties of the sequence tend to be improved as K increases. This is the only method that produces periods greater than M [24, 25]
Its recursive equation is:

$$X_{i+1} = X_n + X_{n-k} \mod (m) \qquad i = n+1, n+2, n+3, \ldots, N$$

The numbers r_i can be generated by the equation:

$$r_i = \frac{X_i}{m-1}$$

Exercise 1. Generate numbers pseudo congruence additive.
Generate 7 pseudo-random numbers from the following data with M = 100.

$$X_i = X_{i-1} + X_{i-n} \mod (m)$$

$$r_i = \frac{X_i}{m-1}$$

x1	x2	x3	x4	x5
65	89	98	3	69

$$\begin{array}{ll}
x_1 = (69 * 65) \mod (100) = 34 & r_1 = \frac{34}{99} = 0.34 \\
x_2 = (34 * 89) \mod (100) = 23 & r_2 = \frac{23}{99} = 0.23 \\
x_3 = (23 * 98) \mod (100) = 21 & r_3 = \frac{21}{99} = 0.21 \\
x_4 = (21 * 3) \mod (100) = 24 & r_4 = \frac{24}{99} = 0.24 \\
x_5 = (24 * 69) \mod (100) = 93 & r_5 = \frac{93}{99} = 0.24 \\
x_6 = (93 * 34) \mod (100) = 27 & r_6 = \frac{27}{99} = 0.27 \\
x_7 = (27 * 23) \mod (100) = 50 & r_7 = \frac{50}{99} = 0.5
\end{array}$$

Exercise 2. Generate numbers pseudo congruence additive.
Generate 10 pseudo numbers from the following data with M = 86.

$$X_i = X_{i-1} + X_{i-n} \mod (m)$$

$$r_i = \frac{X_i}{m-1}$$

x1	x2	x3	x4	x5
3	17	73	31	42

$x_1 = (42 + 3) \bmod (86) = 45 \quad r_1 = \frac{45}{85} = 0.5294$
$x_2 = (45 + 17) \bmod (86) = 62 \quad r_2 = \frac{62}{85} = 0.7294$
$x_3 = (62 + 73) \bmod (86) = 49 \quad r_3 = \frac{49}{85} = 0.5767$
$x_4 = (49 + 31) \bmod (86) = 80 \quad r_4 = \frac{80}{85} = 0.9411$
$x_5 = (80 + 42) \bmod (86) = 36 \quad r_5 = \frac{36}{85} = 0.4235$
$x_6 = (36 + 45) \bmod (86) = 81 \quad r_6 = \frac{81}{85} = 0.9829$
$x_7 = (81 + 62) \bmod (86) = 57 \quad r_7 = \frac{57}{85} = 0.6705$
$x_8 = (57 + 49) \bmod (86) = 20 \quad r_8 = \frac{20}{85} = 0.2352$
$x_9 = (20 + 80) \bmod (86) = 14 \quad r_9 = \frac{14}{85} = 0.1647$
$x_{10} = (14 + 36) \bmod (86) = 50 \quad r_1 0 = \frac{50}{85} = 0.5882$

Exercise 3. Generate numbers pseudo congruence additive.
Generate 10 random numbers from the integer sequence with M = 10.

$$X_i = X_{i-1} + X_{i-n} \bmod (m)$$

$$r_i = \frac{X_i}{m-1}$$

x1	x2	x3	x4	x5
2	9	4	6	8

$x_1 = (8 + 2) \bmod (10)$
$x_1 = 10 \bmod (10)$
$x_1 = 1$
$x_1 = 0.0 * 10$
$x_1 = 0$

$$r_1 = \frac{0}{10-1}$$

$$r_1 = \frac{0}{9}$$

$$r_1 = 0$$

$x_2 = (0 + 9) \bmod (10)$
$x_2 = 9 \bmod (10)$
$x_2 = 0.9$

$x_2 = 0.9 * 10$
$x_2 = 9$

$$r_2 = \frac{9}{10 - 1}$$

$$r_2 = \frac{9}{9}$$

$$r_2 = 1$$

$x_3 = (9 + 4) \bmod (10)$
$x_3 = 13 \bmod (10)$
$x_3 = 1, 3$
$x_3 = 0.3 * 10$
$x_3 = 3$

$$r_3 = \frac{3}{10 - 1}$$

$$r_3 = \frac{3}{9}$$

$$r_3 = 0.333$$

$x_4 = (3 + 6) \bmod (10)$
$x_4 = 9 \bmod (10)$
$x_4 = 0.9$
$x_4 = 0.9 * 10$
$x_4 = 9$

$$r_4 = \frac{9}{10 - 1}$$

$$r_4 = \frac{9}{9}$$

$$r_4 = 1$$

Exercise 4. Generate numbers pseudo congruence additive.
Generate 9 random numbers from the integer sequence with m = 7.

$$X_i = X_{i-1} + X_{i-n} \bmod (m)$$

$$r_i = \frac{X_i}{m - 1}$$

x1	x2	x3	x4	x5
12	5	9	11	4

Exercise 5. Generate numbers pseudo congruence additive.
Generate 12 random numbers from the integer sequence with M = 210.

$$X_i = X_{i-1} + X_{i-n} \bmod (m)$$

$$r_i = \frac{X_i}{m-1}$$

x1	x2	x3	x4	x5
23	22	67	32	45

Exercise 6. Generate numbers pseudo congruence additive.
Generate 10 random numbers from the integer sequence with M = 110.

$$X_i = X_{i-1} + X_{i-n} \bmod (m)$$

$$r_i = \frac{X_i}{m-1}$$

x1	x2	x3	x4	x5
92	94	96	98	99

Exercise 7. Generate numbers pseudo congruence additive.
Generate 20 random numbers from the integer sequence with M = 320.

$$X_i = X_{i-1} + X_{i-n} \bmod (m)$$

$$r_i = \frac{X_i}{m-1}$$

x1	x2	x3	x4	x5
123	435	678	567	129

3.6 Proposed Problems

Determine the period of the following mixed congruence generators:
$X_{n+1} = (8X_n + 16) \bmod 100$ and $X_0 = 15$.
$X_{n+1} = (50\ X_n + 17) \bmod 64$ and $X_0 = 13$.
$X_{n+1} = (5\ X_n + 24) \bmod 32$ and $X_0 = 7$.
$X_{n+1} = (5\ X_n + 21) \bmod 100$ and $X_0 = 3$.
$X_{n+1} = (9\ X_n + 13) \bmod 32$ and $X_0 = 8$.

Determine the period of the following multiplicative congruence generators:
$X_{n+1} = 203\ X_n \bmod 105$ and $X_0 = 17$.
$X_{n+1} = 211\ X_n \bmod 108$ and $X_0 = 19$.
$X_{n+1} = 221\ X_n \bmod 103$ and $X_0 = 3$.
$X_{n+1} = 5\ X_n \bmod 64$ and $X_0 = 7$.
$X_{n+1} = 11\ X_n \bmod 128$ and $X_0 = 9$.

Generate random numbers between 0 and 1 with the following congruence generators and determine the life cycle of each one.
$X_{n+1} = (40\ X_n + 13) \bmod 33$ and $X_0 = 302$.
$X_{n+1} = (71\ X_n + 57) \bmod 341$ and $X_0 = 71$.
$X_{n+1} = (71\ X_n + 517) \bmod 111$ and $X_0 = 171$.
$X_{n+1} = (71561\ X_n + 56822117) \bmod 341157$ and $X_0 = 31767$.
$X_{n+1} = (723\ X_n + 531) \bmod 314$ and $X_0 = 927$.
$X_{n+1} = (452\ X_n + 37452) \bmod 1231$ and $X_0 = 4571$.
$X_{n+1} = (17\ X_n) \bmod 37$ and $X_0 = 51$.
$X_{n+1} = (16\ X_n + 4) \bmod 14$ and $X_0 = 22$.

Generate 15 random numbers if the first 5 sequence numbers are:
1, 13, 17, 12, 4 with $m = 19$
6, 2, 7, 15, 5 with $m = 24$
14, 11, 10, 16, 18 with $m = 30$
40, 61, 81, 4, 25 with $m = 100$
987, 173, 451, 438, 611 with mod 1000
86, 95, 110, 73, 91 with m 500
99, 61, 118, 142, 95 with m 320

References

1. Ríos Insua, David, Sixto Ríos Insúa, Jacinto Martín Jiménez, and Antonio Jiménez Martín. 2009. *Simulación: métodos y aplicaciones*. Number 004.94. Alfaomega.
2. Mancilla Herrera, Alfonso Manuel. 2011. Números aleatorios historia, teoría y aplicaciones. *Revista Científica Ingeniería y Desarrollo* 8: 49–69.

3. López María Victoria. 2005. Software para la generación de variables aleatorias empleadas en simulación. In *VII Workshop de Investigadores en Ciencias de la Computación*.
4. Barroso, Elina Miret, Gladys Linares Fleites, and María V Mederos Bru. 2002. *Estudio comparativo de procedimientos de escalamiento multidimensional a través de procedimientos de escalamiento multidimensional a través de experimentos de simulación*. Editorial Universitaria.
5. Ochoa, JC. 2002. Modelo estocástico de redes neuronales para la síntesis de caudales aplicados a la gestión probabilística de sequías. Ph.D. thesis, Tesis Doctoral. Escuela Técnica Superior de Ingenieros de Caminos Canales y...
6. Bello, L, and M Bertacchini. 2009. Generador de números pseudo-aleatorios predecible en debian. In *III International cyber security conference, Manizales, Colombia*.
7. Primorac, Carlos R, María V López, and Sonia I Marino. 2011. Construcción de una librería de números pseudoaleatorios y muestras artificiales con matlab. *Revista de la Escuela de Perfeccionamiento en Investigación Operativa* 19 (32): 241–257.
8. López, María Victoria, and Sonia Itatí Mariño. 2003. Pruebas de hipótesis para generadores de números pseudoaleatorios. In *IX Congreso Argentino de Ciencias de la Computación*.
9. Solano, Humberto Llinás, and Carlos Rojas Álvarez. 2005. *Estadística descriptiva y distribuciones de probabilidad*. Universidad del Norte.
10. Kelber, Kristina. 2000. N-dimensional uniform probability distribution in nonlinear autoregressive filter structures. *IEEE Transactions on Circuits and Systems I: Fundamental Theory and Applications* 47 (9): 1413–1417.
11. Jerrum, Mark R, Leslie G Valiant, and Vijay V Vazirani. 1986. Random generation of combinatorial structures from a uniform distribution. *Theoretical Computer Science* 43: 169–188.
12. Blum, Lenore, Manuel Blum, and Mike Shub. 1986. A simple unpredictable pseudo-random number generator. *SIAM Journal on Computing* 15 (2): 364–383.
13. Yuste, Santos Bravo, and Héctor Sánchez-Pajares. 2011. Una función random poco aleatoria. *Revista Española de Física* 16 (2).
14. Wichmann, Brian A, and I.D. Hill. 2006. Generating good pseudo-random numbers. *Computational Statistics & Data Analysis* 51 (3): 1614–1622.
15. Butto Zarzar, Cristianne, Joaquín Delgado, and Jerónimo Zamora. 2003. Ejemplos del uso de la hoja de cálculo como herramienta didáctica. *Educación matemática* 15 (3).
16. Sanabria Brenes, Giovanni, and Félix Núñez Vanegas. 2012. Simulación en excel: Buscando la probabilidad de un evento. *Revista Digital: Matemática, Educación e Internet* 12 (2).
17. Périssé, Marcelo Claudio. 2007. Control de la calidad utilizando excel.
18. Pérez, Cecilia, Sonia Itatí Mariño, and María Victoria López. 2011. Desarrollo de generadores de números pseudoaleaorios en octave. *TE & ET* 6: 24.
19. Niederreiter, Harald. 1985. The serial test for pseudo-random numbers generated by the linear congruential method. *Numerische Mathematik* 46 (1): 51–68.
20. Eichenauer-Herrmann, Jürgen, and Harald Niederreiter. 1991. On the discrepancy of quadratic congruential pseudorandom numbers. *Journal of Computational and Applied Mathematics* 34 (2): 243–249.
21. Rodriguez, L. 2011. Simulación, método de montecarlo. Métodos Cuantitativos Organización Industrial.
22. Abad, Ricardo Cao. 2002. *Introducción a la simulación ya la teoría de colas*. Netbiblo.
23. Gimenez Palomares, Fernando. 2018. Método congruencial para la generación de números seudoaleatorios.
24. Dehaene, Stanislas. 1996. The organization of brain activations in number comparison: event-related potentials and the additive-factors method. *Journal of Cognitive Neuroscience* 8 (1): 47–68.
25. Zhang, Yan, Di Fan, and Yan Zheng. 2016. Comparative study on combined co-pyrolysis/gasification of walnut shell and bituminous coal by conventional and congruent-mass thermogravimetric analysis (tga) methods. *Bioresource Technology* 199: 382–385.

Chapter 4
Random Variable Generation Methods

Simulation is the process of designing and developing a computerized model of a system or process and conducting experiments. To understand the system behavior or to evaluate several strategies which the system can be operated based on probabilistic models that allow to generate random variables and obtain significant results through methods such as inverse transform, accept-reject, composition and convolution methods.

4.1 Introduction

The generation of random variables is a process that confronts the simulation because it has variables that have a probabilistic behavior, and where the variability could be classified within some known probability distribution. The generation of any random variable is based on the generation of a uniform distribution (0, 1), and the transformations of these generated numbers in values of other distributions [1–4].

There are usually several algorithms that can be used to generate values from a given distribution, and different factors that can be considered to determine what algorithm to use in a particular case, though these factors often conflict with one another [5, 6]. Some of these factors are as follows:

- **Accuracy**. Values must be obtained from a variable with a given precision. Sometimes, it has enough with getting an approximation and some not.
- **Efficiency**. The algorithm that implements the build method has associated runtime and a memory expense. We will choose a method that is efficient in time and the amount of memory required.
- **Complexity**. We look for methods that have minimal complexity, as long as certain accuracy is guaranteed.

L. Cevallos-Torres and M. Botto-Tobar, *Problem-Based Learning: A Didactic Strategy in the Teaching of System Simulation*, Studies in Computational Intelligence 824, https://doi.org/10.1007/978-3-030-13393-1_4

4.1.1 Types of Random Variables

Once it is clear what exactly is a random variable, the next thing is to define the types of variables that are counted to experiment; this is how the topics of discrete and continuous random variables will be addressed. For the generation of discrete or continuous random variables, it is necessary to have the specific information of the desired distribution, as well as the application of a method for the generation of the random variable, and the computational implementation for Used in the simulation [7, 8].

With what was said in the preceding paragraph, we reached the first type of random variable, called discrete. When experimenting, we are generally interested in some function of the result rather than the result itself. Thus, for example, when throwing a dice twice we could be interested only in the sum of the points obtained and not in the pair of values that gave rise to that value of the sum [3, 4].

In addition to the discrete, there is also another type of random variable called continuous. In this case, the range we work with will cover all the real numbers that exist. When a continuous random variable is used, it works from specific quantities that can be accessed within our experiment. For example, the duration of a call. The operating time of an industrial equipment, the time of repair of a machine, etc. Therefore, for each of these examples, we will be able to set a specified time interval either between 1 min to 5 min or from 1 h up to 8 h, the important thing is to identify that time range there are infinite real numbers [8].

4.1.2 Methods for Generating Random Variables

The generation of any random variable is based on the previous generation of a uniform distribution (0, 1). And, the transformations of these generated numbers in values from other distributions. Most of the techniques used for the generation can be grouped into:

- Inverse transform method.
- Acceptance-rejection method
- Composition method.
- Convolution method.

4.1.3 Inverse Transform Method

This method is applied to the accumulated distribution $F(x)$, from the probability distribution $f(x)$, which will be simulated either by a summation, if it is a discrete variable or through an integration if it is a continuous variable [9, 10]. Since $F(x)$ is between the interval (0, 1), a uniform random number can be generated R_i, to determine the value of the random variable whose accumulated distribution is the same, R_i, which leads solving the following equation:

$$F(x) = R_i \tag{4.1}$$

$$x = F^{-1}(R_i) \tag{4.2}$$

4.2 Inverse Transform and Discrete Random Variables

If we have a discrete variable X, it takes values $x - 1$ with probabilities $P - 1$. The sum is equal to 1, an algorithm to simulate X would be:

Generate values of a uniform U variable (0, 1) and make $X = x_1$, if $U \leq P_i$, and do $X = x_j$ then $\sum_{i=0}^{j-1} p_i < U \leq \sum_{i=0}^{j} p_i$

This method consists of the following steps:

- Define the f density function (x) that represents the variable to the model.
- Calculate the accumulated function $F(x)$.
- Clear the random variable x and obtain the inverse cumulative function, $F(X)^{-1}$.
- Generate random variables x, substituting values with numbers pseudo ri U (0, 1) in inverse cumulative function.

The inverse transform method can be used to simulate discrete random variables, such as Poisson, Bernoulli, binomial, geometric, general discrete. The generation is carried out through the accumulated probability $P(x)$ and the production of pseudo numbers r_i U (0, 1).

4.2.1 Bernoulli Probability Distribution

For this type of distribution only two results are possible: success or failure; where the possibility of *success* is defined as p and *failure* $1 - p$; with this argument we can build a probability function of Bernoulli type (Fig. 4.1), as follows [11, 12]:

$$p(x) = p^x(1 - p)^{1-x}$$

where $x = 0, 1$. The probabilities are calculated to $x = 0$ and $x = 1$, to get:

x	0	1
$p(x)$	$1 - p$	p

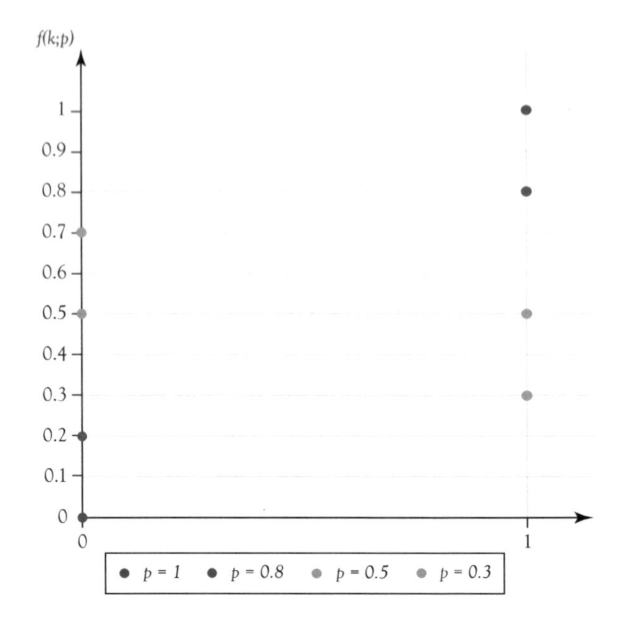

Fig. 4.1 Bernoulli density function graph [13]

Where its distribution function is given by:

$$p(x) = \begin{cases} \text{1-p} & \text{if } x = 0 \\ \text{p} & \text{if } x = 1 \end{cases}$$

Accumulating the $P(x)$ values, it obtains:

x	0	1
$p(x)$	$1-p$	p

$$F(x) = \int_0^1 a - p \qquad dt = 1 - p(t)^{\frac{0}{1}} = 1 - p(1 - 0)$$

$$F(x) = 1 - p \qquad \text{If} \quad 0 \leq x$$

$$1 - x = R_i$$

$$x = 1 - R_i$$

Generating pseudo numbers $r_i \quad U$ (0, 1), it applies the rule:

$$f(x) = \begin{cases} \text{if } r_i(0, 1 - p); & x = 0 \\ r_i(1 - p, 1) & x = 1 \end{cases}$$

Exercise. Below table shows the daily demand for mobile headphones sold in "Mi comisariato" supermarkets [14].

Day	1	2	3	4	5	6	7	8	9
Demand	1	2	2	1	3	0	3	1	3

Simulate the demand behavior using a inverse transform method. From the historical information, the punctual and accumulated probabilities are calculated for $x = 0, 1, 2, 3$ (see below table).

x	$p(x)$	$P(x)$
0	0.1111	0.1111
1	0.2222	0.3333
2	0.3333	0.6666
3	0.3333	1

The rule to generate this random variable would be given by:

$$f(x) = \begin{cases} \text{if } r_i(0 - 0.1111); & x = 0 \\ r_i(0.1111 - 0.3333); & x = 1 \\ r_i(0.3333 - 0.6666); & x = 2 \\ r_i(0.6666 - 1); & x = 3 \end{cases}$$

With the list of pseudo numbers r_i $U(0, 1)$, and the previous rule, it is possible to simulate the daily demand for mobile headphones, as it is shown in below table.

Day	r_i	Daily demand
1	0.213	1
2	0.345	2
3	0.021	0
4	0.987	3
5	0.543	2

4.2.2 *Variables That Follow a Binomial Distribution*

Since the binomial distribution is the sum of n Bernoulli random variables; then the generation of random numbers that follow a binomial distribution implies adding the simulated values of n Bernoulli random variables [15, 16] (see Fig. 4.2). The procedure is as follows:

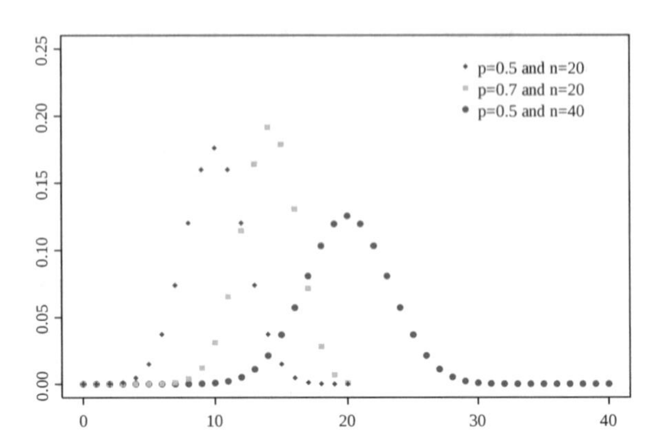

Fig. 4.2 Binomial density function graph [17]

1. Generate n uniform numbers R_i.
2. Count how many of these generated numbers are less than p.
3. The quantity found in step 2 is the simulated value of the random variable x.
4. Repeat the above steps as many times as you want.

Its distribution $f(x)$ is:

$$f(x) = \begin{cases} \frac{n}{x} p^x (1-p)^{n-x}; & \forall x \in \{0, 1, \ldots, n\} \\ 0 & \text{otherwise} \end{cases}$$

where:
n It is the sample size.
x The integer value that takes the random variable: 0,1, 2,…,n.
p It is the probability of success.
$1 - p$ It is the probability of failure.
Its accumulated function $F(x)$ is:

$$F(x) = \int_0^n \frac{n}{x} p^x (1-p)^{n-x} dt$$

Changing variable:

$$F(x) = \sum_{x}^{i=0} x \frac{x}{1} p^i (1-p)^i$$

$$x = \sum_{x}^{i=0} \frac{Ri}{\frac{20}{i} \frac{30}{16}^{2i}}$$

Exercise. If the number of products a customer selects from a specific perch in a small supermarket follows a Binomial distribution with $n = 15$ and $p = 0.4$, simulate purchases from 10 customers and determine:

- What is the total number of products purchased by the 10 customers?
- What is the average of products that were selected?
- What was the minimum and maximum number of products that were selected by a specific customer?

4.2.3 *Variables That Follow a Discrete Uniform Distribution*

It is considered an arbitrary random experiment, and a numerical characteristic is observed in such a way that the result of a random experience can be a finite set of n possible outcomes, all equally probable [18] (see Fig. 4.3). Its distribution $f(x)$ is:

$$f(x) = \frac{1}{j+i+1} \qquad \text{if } \in i, i+1, \ldots, j$$

Its accumulated function, $F(x)$ is:

$$F(x) = \int_i^x \frac{1}{j-i+1} dt = \frac{1}{j-i+1}[x-1] = \frac{x-1}{j-i+1}$$

$$F(x) = \frac{x-i}{j-i+1}$$

$$\frac{x-i}{j-i+1} = R_i$$

$$x = R_i(j-i+1) + i$$

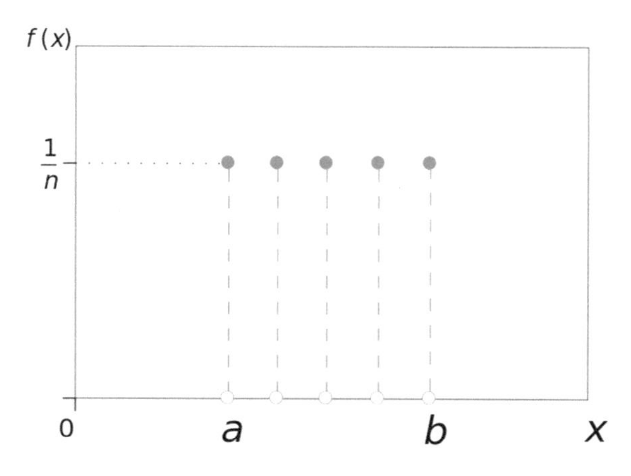

Fig. 4.3 Discrete uniform density function graph [19]

4.2.4 Variables That Follow a Poisson Distribution

The Poisson distribution is used to describe certain types of processes, including number of telephone calls arriving at a commutator, request numbers from patients requiring services in a health institution, car and trucks arrival to a toll, and the accident number registered at certain intersections [20–22] (see Fig. 4.4).

Its distribution $f(x)$ is:

$$p(x) = \begin{cases} \frac{e^{-\lambda}\lambda^x}{x!}; & \text{if} \quad x \in 0, 1, \ldots \\ 0 & \text{otherwise} \end{cases}$$

Its accumulated function, $F(x)$ is:

$$F(x) = \int_0^x \frac{e^{-\lambda}\lambda^x}{x!} dt$$

$$F(x) = e^{-\lambda} \sum x_{i=0} \frac{\lambda^i}{i!} + c$$

$$F(x) = e^{-8} \sum x_{i=0} \frac{8^i}{i!}$$

$$e^{-8} \sum x_{i=0} \frac{8^i}{i!} = R_i$$

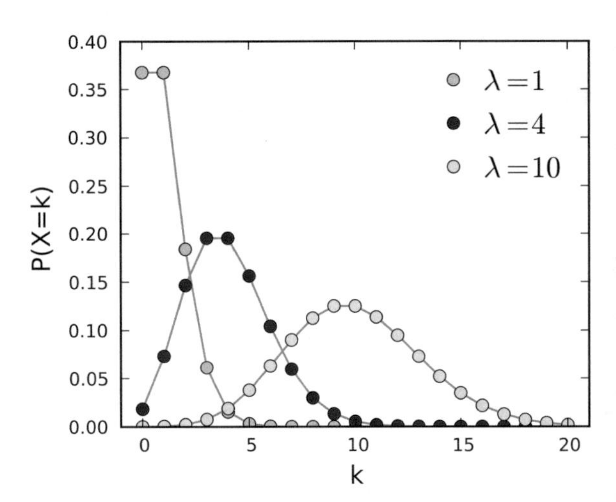

Fig. 4.4 Poisson density function graph [23]

4.2.5 Calculation of the Poisson Distribution

The Poisson probability distribution has to do with certain processes that can be described by a discrete random variable. Generally, the letter X represents this discrete variable and can take values (0, 1, 2, 3, 4, 5…). Where X represents the random variable and the x represents a specific value that that variable can take [24].

Poisson distribution is calculated with the following formula.

$$P(x) = \frac{\lambda^x e^{-\lambda}}{x!}$$

Let's suppose we are re investigating the safety of a dangerous intersection. The records of the Comisión de Tránsito del Guayas (CTG) indicate that on average there are five monthly accidents at the intersection of Víctor Manuel Rendon and Córdova. The number of crashes is distributed according to a Poisson distribution. Thus, the CTG wants to calculate the probability that any month will occur exactly 0, 1, 2, 3, 4 accidents.

Applying the Poisson distribution formula, we have:

$$P(x) = \frac{\lambda^x e^{-\lambda}}{x!}$$

Now, we calculate the probabilities, taking into account that it is a discrete distribution where:
$\lambda = 5$
$x = 0, 1, 2, 3, 4$

The probability that crashes will not occur at this intersection is given by:

$$P(0) = \frac{5^0 e^{-5}}{0!}$$

$$P(0) = \frac{1 * 0.00674}{1}$$

$$P(0) = 0.00674$$

The probability of precisely a crash occurs:

$$P(1) = \frac{5^1 e^{-5}}{1!}$$

$$P(1) = \frac{5 * 0.00674}{1}$$

$$P(1) = 0.03370$$

The probability of exactly two crashes occurring:

$$P(2) = \frac{5^2 e^{-5}}{2!}$$

$$P(2) = \frac{25 * 0.00674}{2 * 1}$$

$$P(2) = 0.08425$$

The probability of exactly three crashes occurring:

$$P(3) = \frac{5^3 e^{-5}}{3!}$$

$$P(3) = \frac{125 * 0.00674}{3 * 2 * 1}$$

$$P(3) = 0.14042$$

The probability of exactly four crashes occurring:

$$P(4) = \frac{5^4 e^{-5}}{4!}$$

$$P(4) = \frac{625 * 0.00674}{4 * 3 * 2 * 1}$$

$$P(4) = 0.17552$$

Our calculations will answer several questions. We may want to know the probability of having 0, 1, 2 monthly crashes. We can figure out this by adding up the probability of having exactly 0, 1 and 2 accidents, as follows:

$$P(0) + P(1) + P(2) = 0.12469$$

To improve the security of the intersection, the CTG has been proposed to take security measures, so they indicate that to do so the number of accidents that occur monthly must exceed 65% ($P(x) = 0.65$).

To solve this problem, you need to calculate the probability of having 0, 1, 2 or 3 crashes and then subtract the result from 1 to get the probability of more than 3 crashes.

$$P(0) + P(1) + P(2) + P(3) = 0.2651$$

As the Poisson probability of three or fewer accidents occurring is 0.26511, and the likelihood of having more than three accidents should be 0.73489 (1– 0.26511). Because 0.73489 is more significant than 0.65, it is necessary to take steps to improve the intersection of the streets. We could continue to calculate the odds for more than four accidents and eventually build a Poisson probability distribution of the number of monthly accidents at this intersection.

Exercise. It knows that the number of customers who arrive every minute to a banking institution can be modeled with a Poisson distribution with average = 7 people/min. It is known that 40% of the customers who arrive at the institution are going to cancel taxes, 20% to make deposits and the rest are directed to the credit bureaus. Simulate 1 h of bank operation and determine the total number of customers who visited each bank section.

4.2.6 *Variables That Follow a Geometric Distribution*

See Fig. 4.5.

$$f(x) = p(1-p)^x \quad \text{if} \quad x \in 0, 1, \ldots$$

$$F(x) = \int_0^x p(1-p)^x dx = 1 - (1-p)^{[x+1]}$$

$$1 - (1-p)^{[x+1]}$$

$$(1-p)^{(x+1)} = 1 - R_i$$

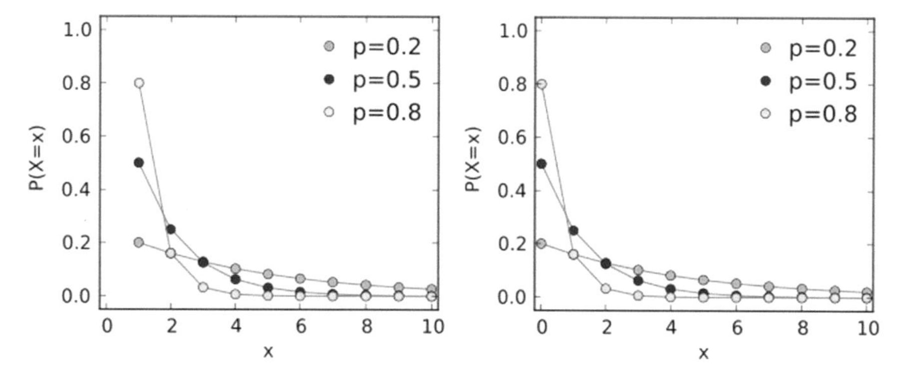

Fig. 4.5 Geometric density function graph [25]

$$(x+1) \quad \log(1-p) = \log 1 - R_i$$

$$x + 1 = \log R_i - \log 0.75$$

$$x = \log R_i - 0.87506$$

4.3 Inverse Transform and Continuous Random Variables

Be F a distribution function (strictly growing) of a continuous random variable X and U a uniform random variable in $(0, 1)$. Thus, $X = f - 1(U)$, is a random variable with distribution F [9, 26].

This method consists of the following steps:

1. Given the probability density function $f(x)$, the cumulative distribution function is developed as:
$$f(x) = \int_{-\infty}^{x} f(t)dt$$
2. A random number is generated $r \in [0, 1]$.
3. $F(x) = R$ is set and the value of X is determined. The variable x is then a continuous random variable of the distribution whose function is given by $f(x)$.

The inverse transform method can also be used to simulate discrete random variables, such as Poisson, Bernoulli, binomial, geometric, general discrete distributions.

4.3.1 *Variables That Follow a Uniform Distribution*

Given a continuous random variable x defined in the interval $[a, b]$ of a real line [27, 28]. In this case, we will say that x has a uniform distribution in the interval $[a, b]$ when its density function is given by [29] (see Fig. 4.6).

$$p(x) = \begin{cases} \frac{1}{b-a}; & a \le x \le b \\ 0 & \text{else} \end{cases}$$

The accumulated function is calculated.

$$F(x) = \int_{a}^{x} \frac{1}{b-a} dt = \frac{1}{b-a} \int_{a}^{x} dt = \frac{1}{b-a}(x-a) = \frac{x-a}{b-a}$$

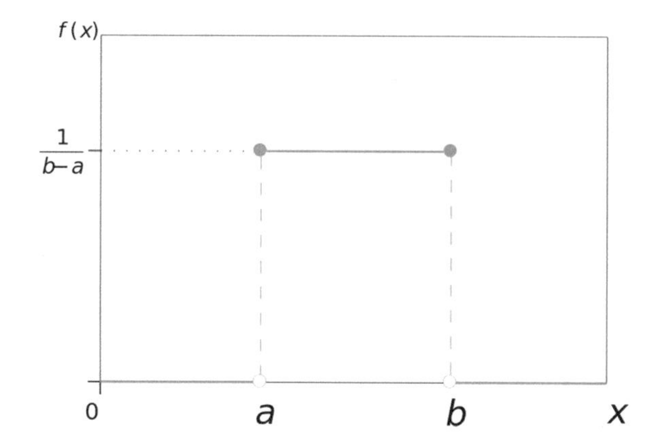

Fig. 4.6 Uniform density function graph [19]

$$F(x) = \begin{cases} 0 & x < a \\ \frac{x-a}{b-a}; & a \leq x \leq b \\ 1 & x > b \end{cases}$$

The inverse function of each segment is cleared, and the r intervals are found.

$$F(x) = r$$

$$\frac{x-a}{b-a} = r$$

$$x - a = r(b - a)$$

$$x = a + r(b - a)$$

$$[h]F(x) = \begin{cases} a + r(b-a) & 0 \leq r \leq 1 \\ 0 & Otherwise \end{cases}$$

Exercise. Find x for $r = 0.4764$; 0.8416; 0.9434; 0.3420; 0.6827; with $b = 12$ and $a = 8$.

$$x = 8 + 0.4764(12 - 8)$$

$$x = 8 + 4(0.4764)$$

$$x = 9.9056$$

$$x = 8 + 0.8416(12 - 8)$$

$$x = 8 + 4(0.8416)$$

$$x = 11.3664$$

$$x = 8 + 0.9434(12 - 8)$$

$$x = 8 + 4(0.9434)$$

$$x = 11.7736$$

$$x = 8 + 0.3420(12 - 8)$$

$$x = 8 + 4(0.3420)$$

$$x = 9.368$$

$$x = 8 + 0.6827(12 - 8)$$

$$x = 8 + 4(0.6827)$$

$$x = 10.7308$$

4.3.2 Variables That Follow a Triangular Distribution

It is called triangular distribution, when it is given by three parameters, which represent the minimum value and the maximum value of the random variable, and the value of the point at which the triangle takes its maximum height. In this case the triangle is not necessarily equilateral [30–33] (see Fig. 4.7).

$$f(x) = \begin{cases} \frac{2(x-a)}{(c-a)(b-a)} & \text{if} \quad a \leq x \leq b \\ \frac{2(c-x)}{(c-a)(c-b)} & \text{if} \quad b \leq x \leq c \end{cases}$$

Its accumulated distribution is:

$$F(x) = \int_a^x \frac{2(x-a)}{(c-a)(b-a)} dt = \frac{2(x-a)}{(c-a)(b-a)}(x-a)$$

$$= \frac{2(x-a)^2}{(c-a)(b-a)}$$

$$F(x) = \int_x^c \frac{2(c-x)}{(c-a)(c-b)} dt = \frac{2(c-x)}{(c-a)(c-b)}(c-x)$$

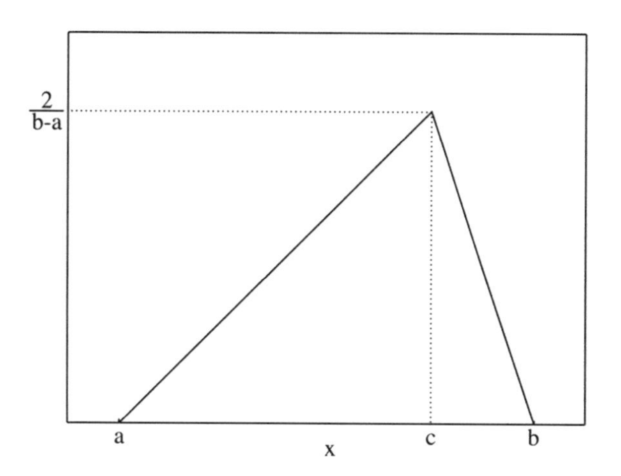

Fig. 4.7 Triangular density function graph [34]

$$= \frac{2(c-x)^2}{(c-a)(c-b)}$$

$$f(x) = \begin{cases} \frac{2(x-a)^2}{(c-a)(b-a)} & \text{if} \quad a \le x \le b \\ \frac{2(c-x)^2}{(c-a)(c-b)} & \text{if} \quad b \le x \le c \end{cases}$$

For the first interval, we have $a \le x \le b$:

$$2(x-a)^2 = R_i(c-a)(b-a)$$

$$(x-a)^2 = \frac{R_i(c-a)(b-a)}{2}$$

$$x = \sqrt{\frac{R_i(c-a)(b-a)}{2}} - a \quad a \le x \le b$$

For the first interval, we have $b \le x \le c$:

$$2(c-x)^2 = R_i(c-a)(b-a)$$

$$(c-x)^2 = \frac{R_i(c-a)(b-a)}{2}$$

$$-x = \sqrt{\frac{R_i(c-a)(b-a)}{2}} - c$$

$$x = c - \sqrt{\frac{R_i(c-a)(b-a)}{2}} \quad \text{if} \quad b \le x \le c$$

Exercise. Given the following function, whose probability distribution $F(x)$ is given by:

$$f(x) = \begin{cases} \frac{1}{2}(x-2) & 2 \le x \le 3 \\ \frac{1}{2}(2-\frac{x}{3}) & 3 \le x \le 6 \\ 0 & Otherwise \end{cases}$$

The accumulated function is calculated.

$$F_1(x) = \int_2^x (t-2)dt = \frac{1}{4}(x-2)^2$$

$$F_2(x) = \int_3^x \frac{1}{2}\left(2 - \frac{t}{3}\right)dt = -\frac{1}{12}(x^2 - 12x + 24)$$

$$F(x) = \begin{cases} \frac{1}{4}(x-2)^2 & 2 \le x \le 3 \\ -\frac{1}{12}(x^2 - 12x + 24) & 3 \le x \le 6 \end{cases}$$

The inverse function of each segment is cleared and the r intervals are found; for the first interval of the function we have:

$F_1(x) = r$
$\frac{1}{4}(x-2)^2 = r$
$x - 2 = \sqrt{4r}$
$x = 2 + \sqrt{4}$

The r-values for the first interval:
$2 = 2 + \sqrt{4r}$
$0 = \sqrt{4r}$
$0 = 4r$
$0 = r$

$3 = 2 + \sqrt{4r}$
$1 = \sqrt{4r}$
$1 = 4r$
$0.25 = r$
For the second function interval we have:
$F_2(x) = r$
$-\frac{1}{12}(x^2 - 12x + 24) = r$
$x^2 - 12x = -24 - 12r$
$x^2 - 12x + 36 = 36 - 24 - 12r$
$(x-6)^2 = 12 - 12r$
$x - 6 = \sqrt{12 - 12r}$
$x = 6 - 2\sqrt{3 - 3r}$

The r-values for the second interval
$3 = 6 - 2\sqrt{3 - 3r}$
$-3 = -2\sqrt{3 - 3r}$
$1.5 = \sqrt{3 - 3r}$
$2.25 = 3 - 3r$
$-0.75 = -3r$
$0.25 = r$

$6 = 6 - 2\sqrt{3 - 3r}$
$0 = -2\sqrt{3 - 3r}$
$0 = \sqrt{3 - 3r}$
$0 = 3 - 3r$
$-3 = -3r$
$1 = r$

$$x = \begin{cases} 2 + \sqrt{4r}; & 0 \le r \le 0.25 \\ 6 - 2\sqrt{3 - 3r}; & 0.25 \le r \le 1 \end{cases}$$

Find x for $r_i = 0.8; 0.63; 0.17$

$x = 6 - 2\sqrt{3 - 3 * 0.8}$
$x = 6 - 2\sqrt{0.6}$
$x = 4.45$

$x = 6 - 2\sqrt{3 - 3 * 0.63}$
$x = 6 - 2\sqrt{1.11}$
$x = 3.89$

$x = 2\sqrt{4 * 0.17}$
$x = 2\sqrt{0.68}$
$x = 2.82$

4.3.3 Variables That Follow an Exponential Distribution

The density function is [35–37] (see Fig. 4.8):

$$f(x) = \begin{cases} \lambda e^{-\lambda} & \ge 0 \\ 0 & x < 0 \end{cases}$$

Its accumulated distribution is:

$$F(x) = 1 - e^{\left(-\frac{x}{\lambda}\right)} \quad \text{if } x \ge 0$$

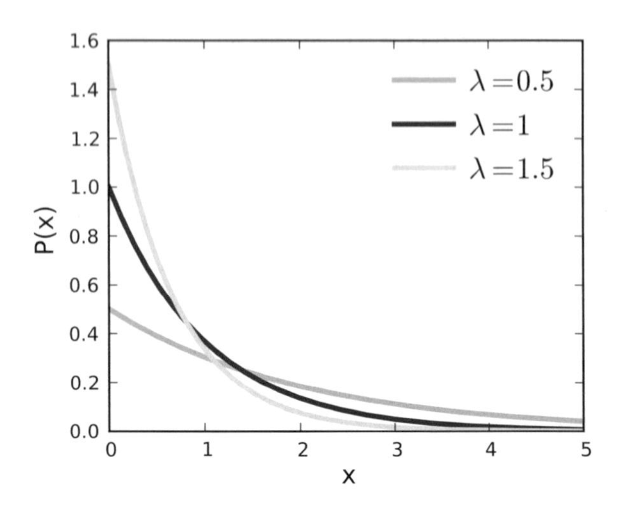

Fig. 4.8 Exponential density function graph [38]

$$F(x) = \int_0^x \lambda e^{-\frac{\lambda}{x}} dx = [e^{-\lambda x}]|\frac{2}{0} = 1 - e^{-\lambda x}$$

$$1 - e^{-\lambda x} = R_i$$

$$lne^{-\lambda x} = R_i$$

$$lne^{-\lambda x} = ln1 - R_i$$

$$-\lambda x lne = ln(1 - R_i)$$

$$x = -\frac{1}{\lambda} ln(1 - R_i)$$

$$x = \begin{cases} -\frac{1}{\lambda} ln(r) & 0 \leq r \leq 1 \\ 0 & otherwise \end{cases}$$

Exercise 1. Find x for r = 0.4466; 0.6427; 0.5902; 0.0318; 0.5901 with $\lambda = 0.01$.
$x = -\frac{1}{0.01} * ln(0.4466)$
$x = -100 * ln(0.4466)$
$x = 80.6$

$x = -\frac{1}{0.01} * ln(0.6427)$
$x = -100 * ln(0.6427)$
$x = 44.2$

$x = -\frac{1}{0.01} * ln(0.5902)$
$x = -100 * ln(0.5902)$
$x = 52.7$

$x = -\frac{1}{0.01} * ln(0.0318)$
$x = -100 * ln(0.0318)$
$x = 344.8$

$x = -\frac{1}{0.01} * ln(0.5901)$
$x = -100 * ln(0.5901)$
$x = 52.7$

Exercise 2. The data of the service time in one of the Invoice collection box of the electric Company of Ecuador behave exponentially with average of 3 minutes/client. A list of numbers pseudo r_i U (0, 1) and the exponential generating equation $x_i = -3ln(1 - r_i)$ allows us to simulate the behavior of the random variable, as shown in below table.

Customer	r_i	Time service (min)
1	0.64	3.06
2	0.83	5.31
3	0.03	0.09
4	0.5	2
5	0.21	0.7

4.3.4 Variables That Follow a Weibull Distribution

The Weibull distribution is a continuous model associated with variables of the type lifetime, failure time, obtaining adequate results of the reliability. Furthermore, this distribution can be defined to include a failure rate or increased or declining risk rate [39–41] (see Fig. 4.9).

The density function is:

$$f(x) = \begin{cases} \sigma\beta^{-\sigma}x^{-\sigma-1}e^{-(\frac{x}{\beta}^{\sigma})} & \text{if } x > 0 \\ 0 & otherwise \end{cases}$$

$$f(x) = \int_0^x \sigma\beta^{-\sigma}x^{\sigma-1}e^{-\frac{x}{\beta}^{\sigma}}\,dy$$

$$f(x) = \int_0^x \sigma\beta^{-\sigma}y^{\sigma-1}e^{-\frac{x}{\beta}^{\sigma}}\,dy$$

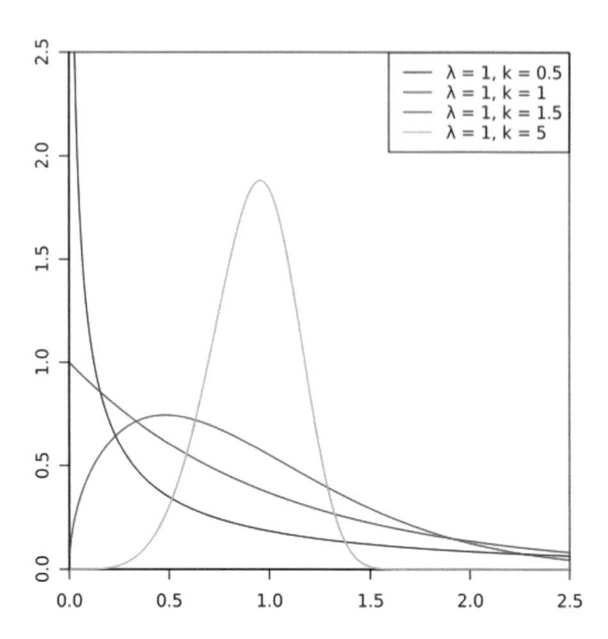

Fig. 4.9 Weibull density function graph [42]

It proceeds to change a variable $2y = (xy)^{\sigma}$.
$f(x) = 1 - e^{-\frac{x}{\beta}^{\sigma}}$
$\sigma = 3; \beta = 1$
$f(x) = 1 - e^{-x^3}$
$1 - e^{-x^3} = R_i$
$e^{-x^3} = 1 - R_i$
$x^3 lne = 1 - R_i$
$x^3 = 1 - R_i$
$x = \sqrt[3]{lnR_i}$

4.4 Inverse Transform and Empirical Distribution

Sometimes the data observed directly will be used to specify a distribution called empirical distribution rather than a theoretical distribution. This will be done when the data does not conform to any known probability distribution [43–46].

A clear disadvantage of the empirical distributions is that the random variables generated during the simulation are never less than X_i or greater than X_n.
Exercise 1.

$$f(x) = \begin{cases} \frac{3}{2}x^2 & -1 \le x \le 1 \\ 0 & otherwise \end{cases}$$

Its accumulated function is:

$$F(x) = \int_{-1}^{2} \frac{3}{2} t^2 dt = \frac{3}{2} \int_{a}^{x} t^2 dt = \frac{1}{2}(x^3 + 1)$$

$$F(x) = \frac{1}{2}(x^3 + 1)$$

The inverse function of each segment is cleared and the r intervals are found.
$F(x) = r$
$\frac{1}{2}(x^3 + 1) = r$
$(x^3 = 2r - 1$
$x = \sqrt[3]{2r - 1}$

$-1 = \sqrt[3]{2r - 1}$
$-1 = 2r - 1$
$0 = 2r - 1$
$0 = r$

$0 = \sqrt[3]{2r - 1}$
$0 = 2r - 1$
$1 = 2r$
$\frac{1}{2} = r$

$1 = \sqrt[3]{2r - 1}$
$1 = 2r - 1$
$2 = 2r$
$1 = r$

$$f(x) = \begin{cases} \sqrt[3]{2r - 1} & 0 \le x \le 1 \\ 0 & otherwise \end{cases}$$

Exercise 2.

$$f(x) = \begin{cases} \frac{1}{4} & 0 \le x \le 1 \\ \frac{3}{4} & 1 \le x \le 2 \end{cases}$$

Its accumulated function is:

$$F_1(x) = \int_{0}^{x} \frac{1}{4} dt = \frac{1}{4} \int_{0}^{x} dt = \frac{x}{4}$$

$$F_1(x) = \frac{x}{4}$$

$$F_2(x) = \int_1^x \frac{1}{4}dt = \frac{1}{4} + \frac{3}{4}\int_0^x dt = \frac{1}{4} + \frac{3x}{4} - \frac{3}{4} = \frac{3x}{4} - \frac{2}{4}$$

$$F_2(x) = \frac{3x - 2}{4}$$

$$f(x) = \begin{cases} \frac{x}{4} & 0 \leq x \leq 1 \\ \frac{3x-2}{4} & 1 \leq x \leq 2 \end{cases}$$

The inverse function of each segment is cleared and the r intervals are found.

$F_1(x) = r$
$\frac{x}{4} = r$
$x = 4r$

$0 = 4r$
$0 = r$

$1 = 4r$
$\frac{1}{4} = r$
$F_2(x) = r$
$\frac{3x-2}{4} = r$
$3x = 4r + 2$
$x = \frac{4r+2}{3}$

$1 = \frac{4r+2}{3}$
$3 = 4r + 2$
$1 = 4r$
$\frac{1}{4} = r$

$2 = \frac{4r+2}{3}$
$6 = 4r + 2$
$4 = 4r$
$1 = r$

$$x = \begin{cases} 4r; & 0 \leq r \leq 0.25 \\ \frac{4r+2}{3} & 0.25 \leq r \leq 2 \end{cases}$$

Find x for $r = 0.8; 0.2; 0.5$.

$x = \frac{4(0.8)+2}{3}$
$x = \frac{5.2}{3}$

$x = 1.73$

$x = 4(0.2)$
$x = 0.8$

$x = \frac{4(0.5)+2}{3}$
$x = \frac{4}{3}$
$x = 1.33$

Exercise 3. Given the following empirical probability distribution, apply the inverse transform method to simulate 100 numbers that follow this probability distribution.

x	0	1	2	3	4	5	6	7	8	9	10	11	12	13	14	15
$f(x)$	0.0067	0.0337	0.0842	0.1404	0.1755	0.1755	0.1462	0.1044	0.0653	0.0363	0.0181	0.0082	0.0034	0.0013	0.0005	0003

4.5 Accept-Reject Method

This method is more probabilistic than the previous one. The methods of investment, composition, and convolution are direct generation methods, in the sense that they deal directly with the distribution function. The accept-reject method is less direct in its approximation [47–49].

In this case, we have the density function $f(x)$ of the variable, and we need a function $t(x)$ that the dimension, i.e., $t(x)^3 f(x)x$. Note that $t(x)$ is not generally a function of density.

$$c = \int_{-\infty}^{+\infty} t(x)dx \geq \int_{-\infty}^{+\infty} f(x)dx = 1$$

but the function $r(x) = \frac{t(x)}{c}$, if it is clearly a function of density. (We assume that t is such that $c <$. We must be able to generate (we hope that easily and quickly) value of the random variable that follows the function $r(x)$. The general algorithm is as follows: Generate x that follows the distribution $r(x)$ generate U $U(0, 1)$ independent of x

$$\text{if } u \leq \frac{f(x)}{t(x)}$$

then return x if you do not repeat the algorithm the algorithm continues to repeat until a value is generated that is accepted.

To make the least number of possible points reject the function $t(x)$ must be the minimum function that dimension to $f(x)$.

4.6 Composition Method

This method will be able to be applied when the density function is easy to

$$f(x) = \sum_{i=1}^{n} t_i x$$

being n the number of pieces in which the function has been divided. Each one of the fragments can be expressed as a product of a distribution function and a weight

$$t_i(x) = f_i(x) w_i$$

and the global distribution function we can get as

$$f(x) = \sum_{i=1}^{n} w_i f_i(x) \text{ with } \sum_{i=1}^{n} w_i = 1.$$

The method is to generate two random numbers, one serves to select one piece, and the other is used to generate a value of a variable that follows the distribution of that piece. The value of the variable obtained is the value sought.

The general algorithm is as follows:
Generate U1, U2 U (0.1)
If u1 = W1 then generate x f1(x)
Yes No
if u1 = w1 + W2 then generate x F2(x)

4.7 Convolution Method

Many random variables including normal, binomial, Poisson, Gamma, Erlang, etc., They can be expressed precisely or approximately by the linear sum of other random variables. The convolution method can be used as long as the random variable x can be expressed as a linear combination of k random variables:

$$x = b_1 x_1 + b_2 x_2 + \cdots + b_n x_n$$

In this method, you need to generate K random numbers $(u1, u2, \ldots, u(k))$ to generate $(u1, u2, \ldots, u(k))$ random variables using any of the above methods to obtain a value of the variable that is desired to get by convolution.

Table 4.1 Daily demand

Daily demand	$p(x)$
0	0.1111
1	0.2222
2	0.3333
3	0.3333

4.8 Proposed Exercises

Find the accumulated and inverse transform of the following distribution functions and find x for the random numbers between 0 and 1 corresponding.

1.
$$f(x) = \begin{cases} \sigma\beta(\beta x)^{\sigma-1} * e^{-(\beta x)^{\sigma}} & x \geq 0 \\ 0 & otherwise \end{cases}$$

   for $R = 0.1562; 0.6184; 0.9553$ with $\sigma = 2$ and $\beta = 3$.
2. $f(x) = \frac{2}{x^3}$; $1 \leq x < \infty$, for $R = 0.7986; 0.8373; 0.4866$.
3. $f(x) = ax^{-(a+1)}$; $1 \leq x < \infty$, for $R = 0.1536; 0.0795; 0.2704$ with $a = 5$.
4. From the following empirical function that determines the probability of the daily demand X of a product with $R = 0213; 0345; 0.021\ 0987; 0543$ for the next 5 days (see Table 4.1).
5. $(x) = \frac{1}{\pi(1+x^2)} 0 \leq x \leq 1$; for $R = 0.0717; 0.3613; 0.5461$.
6.
$$f(x) = \begin{cases} \frac{2(x-a)}{(b-a)(c-a)} & a \leq x \leq c \\ \frac{2}{b-a} & x = c \\ \frac{2(b-x)}{(b-a)(b-c)} & c \leq x \leq b \\ 0 & otherwise \end{cases}$$

   For $a = 12.2$; $b = 35.8$; $c = 19.8$; $R = 0.3961; 0.1223; 0.6101$.

References

1. Mariño, Sonia Itatí, and María Victoria López. 2001. Aprendizaje de muestras artificiales de variables aleatorias discretas asistido por computadora. In *VII Congreso Argentino de Ciencias de la Computación*.
2. Primorac, C., S.I. Mariño, and M.V. López. 2010. Programación en octave de una librería de métodos especiales para generar muestras artificiales de variables aleatorias discretas. ii encuentro regional argentino brasilero de investigación operativa (erabio). In *XXIII Encuentro Nacional de Docentes en Investigación Operativa (ENDIO) y XXI Jornadas de la Escuela de Perfeccionamiento en Investigación Operativa (EPIO)*, 14.

3. Arcay, Alfonso Orro, and Francisco García Benítez. 2006. *Modelos de elección discreta en transportes con coeficientes aleatorios*. Cátedra abertis.
4. Ruiz, B. 2013. *Un Acercamiento Cognitivo y Epistemológico a la Didáctica del Concepto de Variable Aleatoria*. Ph.D. thesis.
5. López, María Victoria. 2005. Software para la generación de variables aleatorias empleadas en simulación. In *VII Workshop de Investigadores en Ciencias de la Computación*.
6. Badii, M.H., and J. Castillo. 2009. Distribuciones probabilísticas de uso común. *Revista Daena (International Journal of Good Conscience)* 4 (1)
7. Sanabria, G. 2013. *Simulación en excel de variables aleatorias continuas, memorias del iii encuentro sobre didáctica de la probabilidad, la estadística y el análisis de datos (iv edepa)*. Costa Rica: Cartago.
8. LópezMaria V., and Sonia I Mariño. 2002. Software interactivo para la enseñanza-aprendizaje de muestras artificiales de variables aleatorias continuas. In *VIII Congreso Argentino de Ciencias de la Computación (CACIC 2002)*. Universidad de Buenos Aires.
9. Reza, Mohammadreza, and E. Garcia. 2015. Metodo de la transformada inversa.
10. Cuadras, Carles M. 2016. *Problemas de probabilidades y estadística. vol. 2. Inferencia estadística*, vol. 2. Edicions Universitat Barcelona.
11. Landro, Alberto, and MIRTA GONZÀLEZ. 2013. Bernoulli, de moivre, bayes, price y los fundamentos de la inferencia inductiva. *Cuadernos del CIMBAGE* (15).
12. Girón, Francisco Javier. 1994. Historia del cálculo de probabilidades: de pascal a laplace. *Historia de la Ciencia Estadistica. Real Acad. Cien. Exac Fis. y Nat. Madrid*.
13. Wikimedia Commons. File:bernoulli distribution pdf.svg—wikimedia commons, the free media repository, 2017. [Online; Accessed 30 Dec 2018].
14. García, Jaime, and Ernesto Sánchez Sánchez. 2013. Niveles de razonamiento probabilístico de estudiantes de bachillerato frente a una situación básica de variable aleatoria y distribución. *Probabilidad Condicionada: Revista de didáctica de la Estadística* 1: 417–424.
15. Landín, Pedro Rubén, and Ernesto Sánchez. 2010. Niveles de razonamiento probabilístico de estudiantes de bachillerato frente a tareas de distribución binomial. *Educação Matemática Pesquisa: Revista do Programa de Estudos Pós-Graduados em Educação Matemática* 12 (3).
16. Sánchez, E., and P.R. Landín. 2011. Fiabilidad de una jerarquía para evaluar el razonamiento probabilístico acerca de la distribución binomial. In *Investigación en Educación Matemática XV*, 533–542.
17. Wikimedia Commons. 2014. File:binomial distribution pmf.svg—wikimedia commons, the free media repository. [Online; Accessed 30 Dec 2018].
18. Kearns, Michael, Yishay Mansour, Dana Ron, Ronitt Rubinfeld, Robert E. Schapire, and Linda Sellie. 1994. On the learnability of discrete distributions. In *Proceedings of the twenty-sixth annual ACM symposium on theory of computing*, 273–282. ACM.
19. Wikimedia Commons. 2017. File:uniform distribution pdf svg.svg — wikimedia commons, the free media repository. [Online; Accessed 30 Dec 2018].
20. Déniz, Emilio Gómez., José María Sarabia Alzaga, and Faustino Prieto Mendoza, et al. 2009. La distribución de poisson-beta: aplicaciones y propiedades en la teoría del riesgo colectivo.
21. Navarro, A., F. Utzet, P. Puig, J. Caminal, and M. Martín. 2001. La distribución binomial negativa frente a la de poisson en el análisis de fenómenos recurrentes. *Gaceta Sanitaria* 15 (5): 447–452.
22. Correa, J.C., and F. Castrillón. 2006. Comparación por intervalos entre diferentes métodos de estimación de la media de la distribución poisson. *Revista EAFIT* 42: 81–96.
23. Wikimedia Commons. File:poisson pmf.svg — wikimedia commons, the free media repository, 2018. [Online; Accessed 30 Dec 2018].
24. Arroyo, Indira, Luis C. Bravo, Humberto Llinás, and Fabián L Muñoz. 2014. Distribuciones poisson y gamma: Una discreta y continua relación. *Prospectiva* 12 (1): 99–107.
25. Wikimedia Commons. 2016. File:geometric pmf.svg — wikimedia commons, the free media repository. [Online; Accessed 30 Dec 2018].
26. Solano, Humberto Llinás, and Carlos Rojas Álvarez. 2005. *Estadística descriptiva y distribuciones de probabilidad*. Universidad del Norte.

27. Marsaglia, George. 1965. Ratios of normal variables and ratios of sums of uniform variables. *Journal of the American Statistical Association* 60 (309): 193–204.
28. Jerrum, Mark R., Leslie G. Valiant, and Vijay.V. Vazirani. 1986. Random generation of combinatorial structures from a uniform distribution. *Theoretical Computer Science* 43: 169–188.
29. Kelber, Kristina. 2000. N-dimensional uniform probability distribution in nonlinear autoregressive filter structures. *IEEE Transactions on Circuits and Systems I: Fundamental Theory and Applications* 47 (9): 1413–1417.
30. Back, W. Edward, Walter W. Boles, and Gary T. Fry. 2000. Defining triangular probability distributions from historical cost data. *Journal of Construction Engineering and Management* 126 (1): 29–37.
31. Guang-yu, Z.H.U. 2009. Meme triangular probability distribution shuffled frog-leaping algorithm [j]. *Computer Integrated Manufacturing Systems* 10: 1979.
32. Johnson, David. 1997. The triangular distribution as a proxy for the beta distribution in risk analysis. *Journal of the Royal Statistical Society: Series D (The Statistician)* 46 (3): 387–398.
33. Kokonendji, C.C., T. Senga Kiessé, and Silvio S. Zocchi. 2007. Discrete triangular distributions and non-parametric estimation for probability mass function. *Journal of Nonparametric Statistics* 19 (6—-8): 241–254.
34. Wikimedia Commons. 2014. File:triangular distribution pmf.png — wikimedia commons, the free media repository. [Online; Accessed 30 Dec 2018].
35. Drăgulescu, Adrian, and Victor M Yakovenko. 2001. Exponential and power-law probability distributions of wealth and income in the united kingdom and the united states. *Physica A: Statistical Mechanics and its Applications* 299 (1–2): 213–221.
36. Krohling, Renato A., and Leandro dos Santos Coelho. 2006. Pso-e: Particle swarm with exponential distribution. In *IEEE Congress on Evolutionary Computation, 2006. CEC 2006.*, 1428–1433. IEEE.
37. Drăgulescu, Adrian, and Victor M. Yakovenko. 2001. Evidence for the exponential distribution of income in the usa. *The European Physical Journal B-Condensed Matter and Complex Systems* 20 (4): 585–589.
38. Wikimedia Commons. File:exponential pdf.svg — wikimedia commons, the free media repository, 2016. [Online; Accessed 30 Dec 2018].
39. Cunha, Luis M., Fernanda A.R. Oliveira, and Jorge C. Oliveira. 1998. Optimal experimental design for estimating the kinetic parameters of processes described by the weibull probability distribution function. *Journal of Food Engineering* 37 (2): 175–191.
40. Khan, Muhammad Shuaib, and Robert King. 2013. Transmuted modified weibull distribution: A generalization of the modified weibull probability distribution. *European Journal of Pure and Applied Mathematics* 6 (1): 66–88.
41. Aryal, Gokarna R., and Chris P. Tsokos. 2011. Transmuted weibull distribution: A generalization of theweibull probability distribution. *European Journal of Pure and Applied Mathematics* 4 (2): 89–102.
42. Wikimedia Commons. 2012. File:weibull pdf.svg — wikimedia commons, the free media repository. [Online; Accessed 30 Dec 2018].
43. Turnbull, Bruce W. 1976. The empirical distribution function with arbitrarily grouped, censored and truncated data. *Journal of the Royal Statistical Society. Series B (Methodological)* 290–295.
44. Loynes, R.M. 1980. The empirical distribution function of residuals from generalised regression. *The Annals of Statistics* 285–298.
45. Feller, William. 2015. On the kolmogorov–smirnov limit theorems for empirical distributions. In *Selected Papers I*, 735–749. Berlin: Springer.
46. Dedecker, Jérôme, Florence Merlevède., and Emmanuel Rio. 2014. Strong approximation of the empirical distribution function for absolutely regular sequences in rd. *Electronic Journal of Probability* 19 (9): 1–56.
47. Kim, Insu. 2017. Markov chain monte carlo and acceptance-rejection algorithms for synthesising short-term variations in the generation output of the photovoltaic system. *IET Renewable Power Generation* 11 (6): 878–888.

48. Martino, Luca, and David Luengo. 2018. Extremely efficient acceptance-rejection method for simulating uncorrelated nakagami fading channels. In *Communications in Statistics-Simulation and Computation* 1–17.
49. Moretón, Rodrigo, Eduardo Lorenzo, and Luis Narvarte. 2015. Experimental observations on hot-spots and derived acceptance/rejection criteria. *Solar energy* 118: 28–40.

Chapter 5
Monte Carlo Simulation Method

The sequential use of random numbers, to sample the values of probability variables, allows obtaining solutions to mathematical problems such as the Monte Carlo method, that allows to model stochastic parameters or deterministic based on random sampling. To justify the use of this method is needed knowing concepts such as the weak law of large numbers and the central boundary theorem.

5.1 Introduction

The Monte Carlo method is a technique of numerical analysis that is based on the use of a sequence of random numbers, with the purpose of sampling the values corresponding to the probabilistic variables of a certain problem. Due to the high number of possible states of the system, it becomes impossible to calculate the average value of all these states, so it is decided to take a sample and estimate those average values, from probability distributions [1–4].

The first component of a Monte Carlo method calculation is numerical sampling of random variables with specific probability density functions. This is done from different techniques, which are useful for generating random values of a x variable distributed in the range $x_{min} \leq x \leq x_{max}$ according to probability density function $p(X)$ [5–9].

Due to the Monte Carlo method is a research and planning tool; It is an artificial sampling technique used to operate complex systems that have random components numerically. Because of the large amounts of data [9], tools are needed to generate random numbers, that is how, thanks to the advancement of technology and programming languages, it is preferable to use pseudo names because, from a seed, a sequence of numbers is always produced Random equal and evenly distributed between 0 and 1 [10–12]. To further explore this topic, the reader is encouraged to review Chap. 3.

L. Cevallos-Torres and M. Botto-Tobar, *Problem-Based Learning: A Didactic Strategy in the Teaching of System Simulation*, Studies in Computational Intelligence 824, https://doi.org/10.1007/978-3-030-13393-1_5

From a generator of pseudo numbers evenly distributed in the interval (0–1), it is possible to build generators with uniform distributions $p(x)$ trough different procedures, such as the inverse transform, the accept-reject and composition methods. Besides, these methods are analyzed more clearly in Chap. 4 [13, 14].

5.2 Monte Carlo Method Justification

The initial justification for the use of Monte Carlo comes from two central theorems of probability and statistics: the weak law of large numbers and the central boundary theorem (or the central boundary theorem).

5.2.1 Weak Law of Large Numbers

To better understand this concept, we will proceed to explain it through the following example: Let's imagine that an experiment consists of throwing a coin in the air an X number of times, half times approximately appears "face". Similarly, if a dice of six same faces are repeatedly thrown, each of the faces leaves about one-sixth of the times the dice was rolled [15–18].

If the coin or the dice is thrown a certain number of times, let's say 10 or 15 times; the indicated approximation can be reduced to half the times for each side of the coin, or to the sixth of the times for each face of the dice. It would not be unthinkable that 10 times as a result, in nine appears face and only one is cross, in the coin, or, when the dice roll 10 or 15 times, it does not appear the 3 on any occasion.

However, it happens that the higher the number of times the dice is thrown, or that the coin is released, the higher the approximation of the relative frequency (the number of times the event is divided by the total number of times the experiment was performed) to the event probability [19].

It is doubtful that if you throw a coin 3,000 times, the three thousand times the same face appears. And more unlikely is the larger the number of times the roll is repeated. Each face tends to come out the same number of times, that is, half the time, which coincides with the probability ½ of each face. In the case of a dice, the frequency tends to approximate to ⅙.

In general, for any event $x_n = x_n(A)$. $x_n \rightarrow p$ for n large enough. If the event "A" is a face coin: $x_n \rightarrow$ ½, and if it is a dice face: $x_n \rightarrow$ ⅙, for n large enough.

This experimental law known as the law of frequency stability has a clear mathematical backing in a group of rigorous theorems that together shape what we call the law of the high numbers.

In other words, the weak law of the high numbers establishes that if $X_1, X_2, X_3, \ldots$ is an infinite succession of independent random variables that have the same expected value μ and variance σ^2, then the average $\bar{X} = \frac{(X_1+\cdots+X_n)}{n}$ converges in probability to μ. In other words, for any positive number ϵ must be fulfilled in:

$$\lim_{n \to \infty} P(|\bar{X}_n - \mu| < \epsilon) = 1$$

Example 1.
"Emelec" is a local football team of Guayaquil, at this season it has won the last nine games; then, by **the law of large numbers**, it is probable that this Sunday will lose against "Liga Universitaria de Quito". In Emelec's case, the fact that it has won nine consecutive matches, it can deduce that it is a strong team and well equipped to win to the next adversary, even if it were "Liga de Quito".
Example 2.
There is a tv show named "Quiero ser millonario" transmitted Ecuadorian TV station "Ecuavisa". The game is to turn a roulette; so far it has given 120 turns without leaving the jackpot (2.000 USD); thus, by the law of large numbers, in the next turn, there is a lot of chance that this prize will come out.

If we analyze it better, we can interpret that there is no relation between the previous spins of roulette and the next turn; at every turn, all the numbers will have the same probability of going out, independently of numbers that have come out so far.
Example 3.
We suppose to repeat for many times the same experiment, like the dice roll: the probability that leaves 1 is equal to $1/6$, that is 16.66%. Now, if we throw more times the dice, implies that the output frequency of number 1 will approach the probability of 16.6%. In analyzing carefully, we can indicate that the law of large numbers is present in this exercise. Because, when conducting the experiment after 20 roll, this frequency can be both 5 and 25%, after 100 roll it is probable that it between 14 and 20%, after 5,000 roll it is difficult that it is not understood between 16 and 17% and then, the more it advances, the more that value will be closer to 16.6%. since the probability that the numbers 2 come out, 3, 4, 5, 6 is always equal to 16.6%, also the output relative frequencies will tend to flatten all about 16.6%, as you advance with the releases of the die, the output frequencies of the 6 numbers They are always close to a value of 16.6%. Figure 5.1 the relative statistics corresponding to a repeated release of dice [17, 20, 21].

5.2.2 *The Central Boundary Theorem*

It tells us that if a sample is large enough (usually when the sample size n exceeds 30), regardless of the sample distribution mean, it will follow approximately a normal distribution. In other words, given any random variable, if we extract samples of n size ($n > 30$) and calculate the sample averages, these averages will follow a normal distribution. Moreover, the average will be the same as that of the variable of interest, and the standard deviation of the sample mean will be approximately the standard error [22–24].

30 releases			1000 releases		
1	4 Times	13,3%	1	171 Times	17,1%
2	5 Times	16,7%	2	164 Times	16,4%
3	8 Times	26,6%	3	169 Times	16,9%
4	6 Times	20%	4	167 Times	16,7%
5	3 Times	10%	5	164 Times	16,4%
6	4 Times	13,3%	6	165 Times	16,5%

Fig. 5.1 Statistics on the dice roll

Table 5.1 Dice roll probabilities

Number	1	2	3	4	5	6
Probability	$^1/_6$	$^1/_6$	$^1/_6$	$^1/_6$	$^1/_6$	$^1/_6$

In other words, the central boundary theorem establishes that if we have a sample with x_1, x_2, and x_n. Random variables independent and identically distributed, each one with mean σ and standard deviation μ, and are defined as $s_n = x_1 + x_2 + \cdots + x_n$. The s_n distribution is normal asymptotically with a normal mean $n\mu$ and variance $n\sigma^2$, independently of the original distribution $x_1, x_2, \ldots, x_n$ [25, 26].

A specific case of the central boundary theorem is the binomial distribution. From $n = 30$, the binomial distribution behaves statistically as a normal; thus, we can apply the appropriate statistical tests for this distribution [27–30].

5.2.2.1 The Central Boundary Theorem Proved with a Dice Roll

The dice roll results follow a uniform distribution. Each one from 1 to 6 has the same probability of "getting out" in an honest pitch with a professional casino dice. The distribution will be like as shown in Table 5.1.

Let's see what happens when we simulate the dice roll a couple of thousand times. A histogram of its frequencies is shown in Fig. 5.2.

The expected result from a total of 2.173 rolls, is that each number appears approximately $^1/_6$, that is 362 times. No matter how many times it is simulated the dice roll, it will always approach a uniform distribution.

Now, let's roll a dice ten times, and we'll determine the average of these rolls (see Table 5.2).

Fig. 5.2 Dice roll

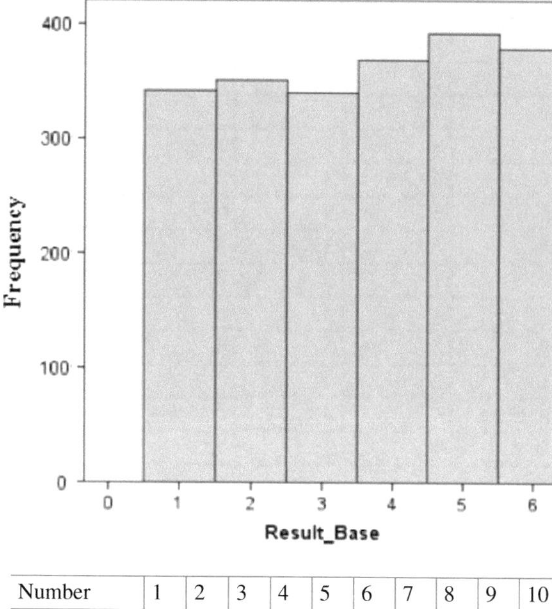

Table 5.2 Dice roll probabilities (10 times)

Number	1	2	3	4	5	6	7	8	9	10
Result	4	3	3	2	3	3	2	4	6	2

And, what happens if we do the same thing several times? Let's say 1,500 (see Table 5.3), and then, we do a histogram its averages.

According to the central boundary theorem, the histogram should look like a normal distribution, as it is seen in the chart in Fig. 5.3.

On the other hand, we have the next question: What will happen if we make a histogram with the sum of the samples of 10? Probably the histogram of n-size sample sums also behaves as a normal distribution, as shown in Fig. 5.4.

5.2.2.2 Exercise Applying Monte Carlo Simulation

Table 5.4 depicts a historical analysis of telephone calls by certain clients in "Claro" call center. From day 1 to day 200; the goal is to measure the number of daily claims that customers make regarding the dissatisfaction n their cable television service.

We can interpret the relative frequency as the probability of occurring an associated event; in this case, the probability of a certain claim numbers (e.g., the probability of three claims given in one day would be 0.30). Thus, Table 5.4 provides the probability distribution associated with a discrete random variable (the random variable is the number of claims to Claro call center, which can only take integer values between 0 and 5).

Let's say we would like to know the expected number (or half) of customer claims per day. To answer this question, we resort to the theory of probability; denoting by

Table 5.3 Dice roll probabilities (1500 times)

Roll	Dice 1	Dice 2	Dice 3	Dice 4	Dice ...	Dice n–1	Dice 1500
1	5	5	3	4	...	6	3
2	5	5	4	3	...	4	2
3	6	1	3	1	...	3	5
4	6	3	5	1	...	3	4
5	5	3	5	1	...	6	5
6	5	3	6	4	...	4	3
7	1	3	3	1	...	2	5
8	4	6	4	4	...	2	3
9	3	2	4	3	...	2	5
10	3	2	3	4	...	2	5
Average	4.3	3.3	4	2.6	...	3.4	4
Sum	43	33	40	26	...	34	40

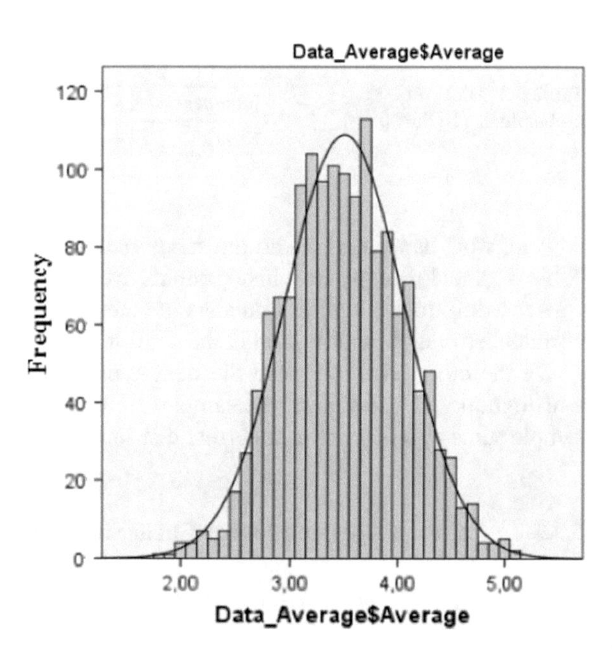

Fig. 5.3 Histogram of 1.500 averages $n = 10$

X to the random variable that represents the daily number of claims in the call center, we know:

$$E[X] = \sum_{i=0}^{5} x_i * P(X = x_i) = 0 * 0.05 + 1 * 0.10 + \cdots + 5 * 0.15 = 2.95$$

where $E[X]$ is the expected value and it is equal to the mean μ.

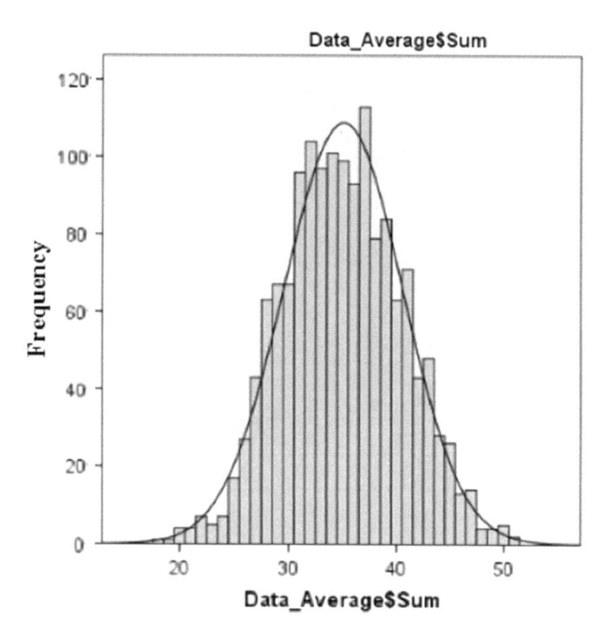

Fig. 5.4 Histogram of 1.500 sums $n = 10$

Table 5.4 Probabilities of claims

Number of claims	Absolute daily frequency	Probability $f(x)$	Cumulated $F(x)$
0	10	0.05	0.05
1	20	0.1	0.15
2	40	0.2	0.35
3	60	0.3	0.65
4	40	0.2	0.85
5	30	0.15	1
Total	200	1	

On the other hand, we can also use Monte Carlo simulation to estimate the expected number of daily claims (in this case it has been possible to obtain the exact value using probability theory; however, it will not always be feasible). Let's see how:

When the probability distribution associated with a discrete random variable is known, it will be possible to use the accumulated relative frequency column to obtain the called "random number intervals associated with each event".

In this case, the intervals obtained are:

- (0.00; 0.05) for the event 0.
- (0.05; 0.15) for the event 1.
- (0.15; 0.35) for the event 2.
- (0.35; 0.65) for the event 3.

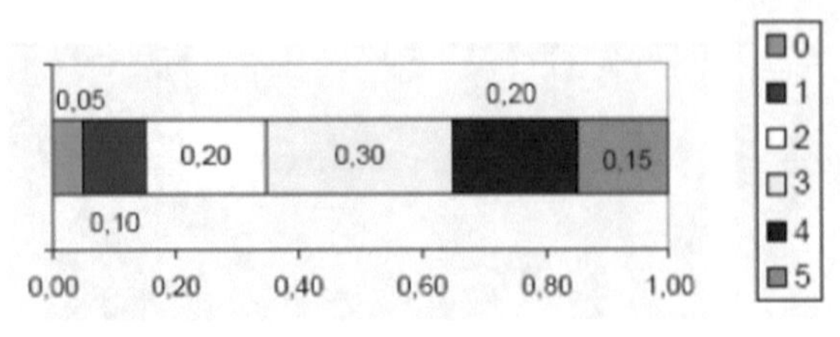

Fig. 5.5 Number of calls

Table 5.5 10 random numbers

Random	R_1	R_2	R_3	R_4	R_5	R_6	R_7	R_8	R_9	R_{10}
R_i	0.3296	0.0899	0.9261	0.9232	0.1485	0.6686	0.3697	0.3368	0.7160	0.8069

Table 5.6 Inverse transform

$F(x)$	$F(x) = R_i$
0.05	$0.00 < x \leq 0.05$
0.15	$0.05 < x \leq 0.15$
0.35	$0.15 < x \leq 0.35$
0.65	$0.35 < x \leq 0.65$
085	$0.65 < x \leq 0.85$
1	$0.85 < x \leq 1.00$

- (0.65; 0.85) for the event 4.
- (0.85; 1.00) for the event 5.

Figure 5.5 shows each of probabilities about the number of customer claims. It depicts the relationship between the probability of each event, and the area that it occupies is appreciated.

Figure 5.5 means, by generating a pseudo-random number with the computer help (from a uniform distribution between 0 and 1); we will be conducting an experiment whose result, obtained randomly and according to the probability distribution previous, it will be associated with an event. Thus, for example, if the computer gives us 0.2567 as number pseudo-random, we can assume that day there have been 2 phone calls from particular clients.

Then, we generate 10 random numbers (see Table 5.5).

Then, we use the inverse transform method (see Table 5.6).

We determine the x value, for the generation of random variables based on the corresponding range and the number R_i, i.e., from these accumulated frequencies can be obtained the intervals of random numbers associated with each operation, each number be linked to a range whose probability is less than or equal to the random number obtained.

x	2	1	5	5	1	4	3	2	4	4

Finally, the average of x values column is calculated:

$$\mu = 2.9999$$

In this case, an estimated value is obtained which corresponds an approximate with the actual value previously calculated via the theoretical definition of the mean. In this case, they are not the same as expected, due to the random component intrinsic to the model, we usually will obtain "close" values to the real value, being those values different from each other (each simulation will provide its results). If instead of using a random sample consisting of 10 observations or hoping that if it had been used 1,000 (or better yet 10,000) observations, the values we would get would be all very close to the actual value corresponding to the expected value, and therefore to the average.

Here, we do not have the number of simulations necessary to be able to conclude, it is instead an example to capture a bit the method. Usually, when we analyze with Monte Carlo, we have 1,000 or 2,000 simulations, as indicated in the preceding paragraph.

In the following chapter of this book, we will use the Monte Carlo simulation, with the different probability distribution functions, these with continuous, discrete or empirical variables.

References

1. Illana, José Ignacio. 2013. Métodos monte carlo. *De Departamento de Física Teórica y del Cosmos, Universidad de Granada* 26 (01).
2. Ximénez, Carmen, and Ana G. García. 2005. Comparación de los métodos de estimación de máxima verosimilitud y mínimos cuadrados no ponderados en el análisis factorial confirmatorio mediante simulación monte carlo. *Psicothema*.
3. Rodrıguez-Aragón, Licesio J. 2011. Simulación, método de montecarlo.
4. Bello, L., and M Bertacchini. 2009. Generador de números pseudo-aleatorios predecible en debian. In *III International Cyber Security Conference*. Manizales, Colombia.
5. Palacios, Simón Pedro Izcara. 2007. *Introducción al muestreo*. Miguel Ángel Porrúa.
6. López, César Pérez. 2000. *Técnicas de muestreo estadístico: teoría, práctica y aplicaciones informáticas*. Alfaomega Grupo Editor.
7. Macías, Alvaro Torres. 1993. *Probabilidad, variables aleatorias, confiabilidad y procesos estocásticos en ingeniería eléctrica*. Uniandes. Fac. de Ingeniería. Depto. de Eléctrica.
8. Gutierrez, Roberto Behar, and Pedro Grima Cintas. 2013. El histograma como un instrumento para la comprensión de las funciones de densidad de probabilidad. In *Probabilidad Condicionada*, 229–235.

9. Moreno, B. 2009. Minería sobre grandes cantidades de datos. *México DF, Universidad Autónoma Metropolitana*, 166.
10. Corrales, Rubén Darío, Daniel St Moran, and Gonzalo Ramírez. 2015. Una aproximación a π con el método de montecarlo mediante el software r: una propuesta para ser llevada al aula de clase. *RECME* 1 (1): 763–766.
11. Vidal, Josep Maria Losilla. 2009. *MonteCarlo toolbox de Matlab: herramientas para un laboratorio de estadística fundamentado en técnicas Monte Carlo*. Universitat Autònoma de Barcelona.
12. Faulín, Javier, and Ángel A. Juan. 2005. Simulación de monte carlo con excel.
13. Alonso, Carlos, and David Ospina. 2001. Un tamaño de muestra preliminar en la estimación de la media, en poblaciones con distribuciones uniformes y triangulares. *Revista Colombiana de Estadística* 24 (1): 27–32.
14. Bergeon, N., S. Kajiwara, and T. Kikuchi. 2000. Atomic force microscope study of stress-induced martensite formation and its reverse transformation in a thermomechanically treated fe-mn-si-cr-ni alloy. *Acta materialia* 48 (16): 4053–4064.
15. Desrosières, Alain. 2004. La política de los grandes números. *Historia de la razón estadística. Barcelona*, Melusina.
16. Jaimes, E., and J Martínez. 2007. *Probability Explorer: Un socio cognitivo en la construcción del significado de la Ley de los Grandes Números con estudiantes de octavo grado en el Instituto Técnico Industrial de Puente Nacional*. Ph.D. thesis, Thesis (Tesis de de especialización en Educación Matemática), Universidad . . ..
17. Yáñez, Gabriel, and Édgar Jaimes. 2013. Efectos de la simulación en la comprensión de la ley de los grandes números. *Integración: Temas de matemáticas* 31 (1): 69–86.
18. Godino, Juan D., Rafael Roa, Angel M. Recio, Francisco Ruiz, and Juan L. Pareja. 2006. Análisis didáctico de un proceso de estudio de la ley empírica de los grandes números. *Statistics (ICOTS 7)* 1.
19. De León, J. 2002. *Estudio de la comprensión de la Ley de los Grandes Números en Estudiantes de nivel Superior: El caso de Ciencias Sociales*. Ph.D. thesis, Tesis de doctorado no publicada, México, Departamento de Matemática . . ..
20. Carvajal, Edgar David Jaimes and Gabriel Yáñez Canal. La ley de los grandes números: Un asunto de experimentación y simulación. *USO DE TECNOLOGÍA EN EDUCACIÓN MATEMÁTICA* 97.
21. Trevethan, H.M., V. YumiKataoka, and M.S. Oliveira. 2010. El uso de juegos para la promoción del razonamiento probabilístico. *Revista Iberoamericana de Educación matemática* 69: 19–33.
22. Alvarado, Hugo, and Carmen Batanero. 2008. Significado del teorema central del límite en textos universitarios de probabilidad y estadística. *Estudios pedagógicos (Valdivia)* 34 (2): 7–28.
23. Alvarado, Hugo, and Carmen Batanero. 2006. El significado del teorema central del límite: evolución histórica a partir de sus campos de problemas. In *Investigación en Didáctica de las Matemáticas/Congreso Internacional sobre Aplicaciones y Desarrollos de la Teoría de las Funciones Semióticas*, 257–277.
24. Martínez, Hugo Alvarado, and Lidia Retamal Pérez. 2012. Dificultades de comprensión del teorema central del límite en estudiantes universitarios. *Educación matemática* 24 (3): 151–171.
25. Arce, Greivin Ramírez, and Kendall Rodríguez. 2018. Simulación de variables aleatorias continuas y el teorema del límite central. *Revista Digital: Matemática, Educación e Internet* 18(1).
26. Díaz, S Pértegas and S Pita Fernández. 2001. La distribución normal. *Cad Aten Primaria* 8: 268–274.
27. Martínez, Hugo Alvarado, and Lidia Retamal Pérez. 2010. La aproximación binomial por la normal: Una experiencia de reflexión sobre la práctica. *Paradígma* 31 (2): 89–108.
28. Retamal, Lidia, Hugo Alvarado, and Rodrigo Rebolledo. 2007. Comprensión de las distribuciones muestrales en un curso de estadística para ingenieros. *Ingeniare. Revista chilena de ingeniería* 15 (1): 6–17.
29. Gómez, Mónica Martínez, and Manuel Daniel Marí Benlloch. 2010. Distribución binomial.
30. Alvarado, Hugo, and Carmen Batanero. 2007. Dificultades de comprensión de la aproximación normal a la distribución binomial. *Números. Revista de Didáctica de las Matemáticas* 67: 1–7.

Chapter 6
Case Study: Logistical Behavior in the Use of Urban Transport Using the Monte Carlo Simulation Method

This study presents a proposal to determine solutions to the models of queue theory through the use of simulation. The main objective is to evaluate the number of people who arrive at a public transport service station in order to be able to minimize monetary losses, the product of the defection of the people of the waiting line of this station. To evaluate the model, we proceeded to use tools that allow simulating random values based on probability distributions; such as the Log-Normal probability distribution, and the Binomial distribution. Our study case was a public taxi transport stop located in Victor Manuel Rendon and Pedro Moncayo streets in Guayaquil city, where it was likely to observe all people waiting for the taxi service. We used simulation methods to obtain estimations from real cases.

6.1 Introduction

The search for explanations for phenomena that happen randomly has led the human being to the need to use mechanisms that allow him to quantify in an imperfect way the possibility of an event occurring. A tool that measures results under some uncertainty event is the probability. To model real situations under probabilistic reasoning, it is essential to use of procedures that allow reasonably to emulate these situations; A resource that helps to obtain estimates or approximations of cases of a real situation is the simulation [1–5].

The construction of a simulation tool has allowed giving feasible and optimal solutions to a problem that is presented daily in a public transport service station, and the citizens very request that at different times of the day. This tool constitutes an alternative to a certain extent economic; for the evaluation of the service quality, and it is useful for supporting decision making, and moreover, allows to define and evaluate common performance measures of waiting for lines, such as the arrival and waiting time of people in the transport station [6–8].

L. Cevallos-Torres and M. Botto-Tobar, *Problem-Based Learning: A Didactic Strategy in the Teaching of System Simulation*, Studies in Computational Intelligence 824, https://doi.org/10.1007/978-3-030-13393-1_6

6.1.1 *Related Work*

Martinez et al. [9] proposed obtaining a quantitative predictive model as an effective way to address the problem of daily demand of passengers in a bus line. In order to solve the problem, the author has used SQL Server Management to tackle and manage simulated data through a database. However, this environment requires users with enough knowledge in this tool.

In the study carried out by Lojano et al. [10] used a hybrid model combining, as they are, the multi-indicator and multiple-cause (MIMIC) model, and the theory of random utility. In order to simulate the demand for passengers, they use the Quito-Cable, the data were obtained by predictions of quantifiable variables such as time, service operational costs, and service prices. However, this research proposed the inverse transformed and Monte Carlo method for obtaining the data; and a computational tool was used for processing them.

6.2 Case Study

A circumstance is given on a daily basis in a taxi's company that transport passenger to Duran city. By using the observation [11, 12] in order to reach this goal, we started by taking data such as the arrival time (time in which people arrive at the transport station), during an hour per day, and also identify the number of people who left the queue for different reasons; this results are marked by "*" and are presented in Table 6.1.

Table 6.1 also presents the time conversion in hours by applying the following formula:

$$conversion = \frac{hour(cell) + minutes(cell)}{\frac{60+seconds(cell)}{3600}} \tag{6.1}$$

Reducing the formula is:

$$conversion = cell * 24 \tag{6.2}$$

6.2.1 *Queue Theory Model*

Queue management or line-of-waiting models consist of some elements that generate waits on entities or agents waiting to be served. In general, queue models have two components. The elements that are processed in the system are entities, and these entities are handled according to a defined criterion [13].

Table 6.1 Arrivals data to the station

	Time	Conversion		Time	Conversion
1	0:00:31	0.00861	26	0:00:40	0.01111
2	0:00:45	0.01250	27	0:01:13	0.02028
3	0:00:27	0.00750	28	0:02:06	0.03500 *
4	0:01:02	0.01722	29	0:01:02	0.01722
5	0:00:50	0.01389	30	0:01:08	0.01889
6	0:01:10	0.01944 *	31	0:00:06	0.00167
7	0:00:59	0.01639	32	0:00:15	0.00417
8	0:00:40	0.01111	33	0:00:34	0.00944 *
9	0:00:52	0.01444 *	34	0:01:02	0.01722
10	0:01:08	0.01889	35	0:01:04	0.01778
11	0:02:03	0.03417	36	0:00:43	0.01194
12	0:00:35	0.00972	37	0:00:19	0.00528 *
13	0:00:27	0.00750	38	0:00:53	0.01472
14	0:00:51	0.01417	39	0:01:32	0.02556
15	0:01:25	0.02361 *	40	0:02:08	0.03556
16	0:01:04	0.01778	41	0:01:21	0.02250
17	0:01:30	0.02500	42	0:01:03	0.01750 *
18	0:01:16	0.02111	43	0:02:02	0.03389
19	0:00:58	0.01611 *	44	0:01:02	0.01722 *
20	0:01:08	0.01889	45	0:01:15	0.02083
21	0:00:44	0.01222	46	0:01:07	0.01861
22	0:00:11	0.00306 *	47	0:00:55	0.01528
23	0:00:17	0.00472	48	0:02:37	0.04361 *
24	0:00:54	0.01500	49	0:00:42	0.01167
25	0:00:37	0.01028	50	0:01:05	0.01806

Two phases were followed for the construction of the simulation model. The first was the identification of the cases to be modeled, and then data time was taken corresponding to the times to arrivals to the system. To adjust the data in a statistical distribution, we proceeded to use stat-fit. that was very helpful to be able to identify which is the probability allocation that best fits our problem [14, 15].

6.2.2 *Using Computer Tools: Stat-Fit*

It is used to analyze and determine the type of probability distribution of a dataset, in such a way that allows to compare the results between several distributions analyzed by a qualification. Once the data have been compiled, an analysis is performed, and then the generated data are shown.

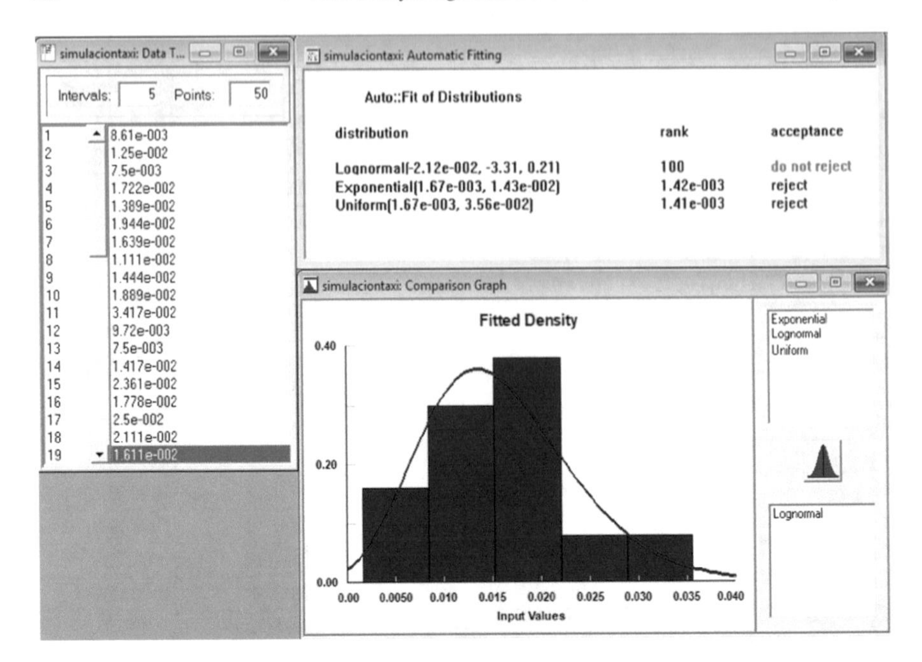

Fig. 6.1 Verification of rejected and non-rejected distributions. Log-normal Distribution Graph

Figure 6.1 shows Stat-Fit probability distributions results, in both distributions that are accepted and those rejected. For this case study, we will use Log-Normal probability distribution [16–18].

6.2.3 *Log-Normal Distribution*

The log-normal distributions was used in [19] to determine arrivals times of buses. Therefore, in this we applied similarly, to simulate the arrival time people to the waiting queue [17].

The log-normal distribution is obtained when the normal distribution describes the logarithms of a variable [20, 21].

Where:

$\mu =$ It is the mean of ln(x).

$\sigma =$ It is the standard deviation of $ln(x)$.

$ln(x) =$ It is a random variable that has a normal distribution.

We built a simulator using a log-normal distribution, to this, the verification process was done that consisted of making the pilot runs and observe the behavior of these.

Then, we determine a procedure that allows creating pseudo values by applying the probability log distribution formula-Normal and the Monte Carlo method [22]. Later, the final results are made the conversion in time format (hh:mm:ss), and it was obtained the probability of each outcome of the arrival people (see Algorithm 1).

Algorithm 1 ResultLognormal()

$Range(\text{"}J25\text{"}).Select$
$ActiveCell.FormulaR1C1 \leftarrow \text{"}N\text{"}$
$Range(\text{"}J2\text{"}).Select$
$ActiveCell.FormulaR1C1 \leftarrow \text{"1"}$
$Range(\text{"}K25\text{"}).Select$
$ActiveCell.FormulaR1C1 \leftarrow \text{"}Random\text{"}$
$Range(\text{"}K76\text{"}).Select$
$ActiveWindow.SmallScrollDown \leftarrow -42$
$Range(\text{"}K26\text{"}).Select$
$ActiveCell.FormulaR1C1 \leftarrow \text{"} = RAND()\text{"}$
$Range(\text{"}L25\text{"}).Select$
$ActiveCell.FormulaR1C1 \leftarrow \text{"}Lognormalformula\text{"}$
$Range(\text{"}L75\text{"}).Select$
$ActiveWindow.SmallScrollDown \leftarrow -33$
$Range(\text{"}L26\text{"}).Select$
$ActiveCell.FormulaR1C1 \leftarrow \text{"}RC[-1] * Lognorm(0.015981, 0.007985,$
$Truncate(0.00167, 0.04361))\text{"}$
$Range(\text{"}M25\text{"}).Select$
$ActiveCell.FormulaR1C1 \leftarrow \text{"}ArrivePERs\text{"}$
$Range(\text{"}M26\text{"}).Select$
$ActiveCell.FormulaR1C1 \leftarrow \text{"} = RC[-1]/24\text{"}$
$Range(\text{"}N25\text{"}).Select$
$ActiveCell.FormulaR1C1 \leftarrow \text{"}PROB\text{"}$
$Range(\text{"}N26\text{"}).Select$
$ActiveCell.FormulaR1C1 \leftarrow \text{"} \leftarrow RC[-1]/R76C13\text{"}$
$Range(\text{"}O25\text{"}).Select$
$ActiveCell.FormulaR1C1 \leftarrow \text{"}ACUM\text{"}$
$Range(\text{"}O75\text{"}).Select$
$ActiveWindow.SmallScrollDown \leftarrow -36$
$Range(\text{"}O26\text{"}).Select$
$ActiveCell.FormulaR1C1 \leftarrow \text{"}RC[-1]\text{"}$
$Range(\text{"}O27\text{"}).Select$

Once the verification is carried out, this behavior is contrasted with the information provided in Table 6.2, where the process of the log-normal distribution is presented using random numbers to generate arrival times for people with their respective probability percentage (Fig. 6.2).

After taking the data, the service times were analyzed with Algorithm 2. We determined the procedure allow to create pseudo values by applying the binomial

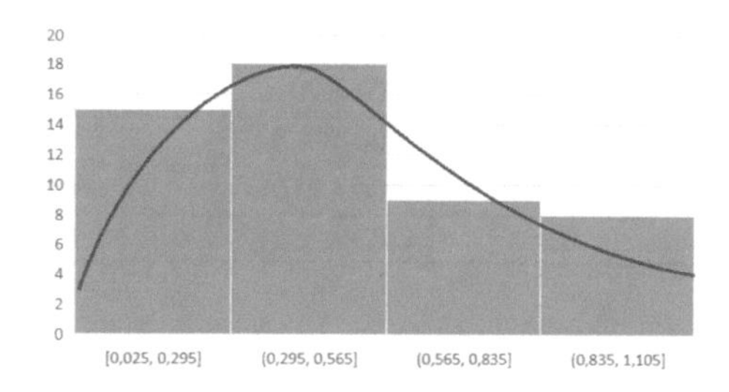

Fig. 6.2 Data generated from the Log-normal distribution

Table 6.2 Data generated from the log-normal distribution

No	Random	Log-normal formula	Arrive PERs	PROB (%)	ACUM (%)
1	0.683	0.00852698	0:00:31	2.38	2
2	0.165	0.00333393	0:00:12	0.93	3
3	0.753	0.00913470	0:00:33	2.55	6
4	0.091	0.00204823	0:00:07	0.57	6
5	0.388	0.00526711	0:00:19	1.47	8
6	0.592	0.00660240	0:00:24	1.84	10
7	0.422	0.00536830	0:00:19	1.50	11
8	0.316	0.00277153	0:00:10	0.77	12
9	0.426	0.00647585	0:00:23	1.81	14
10	0.076	0.00069473	0:00:03	0.19	14
11	0.861	0.01592467	0:00:57	4.44	18
12	0.450	0.00602272	0:00:22	1.68	20
13	0.618	0.00738569	0:00:27	2.06	22
14	0.471	0.00462775	0:00:17	1.29	23
15	0.139	0.00170623	0:00:06	0.48	24
16	0.092	0.00162126	0:00:06	0.45	24
17	1.000	0.01798504	0:01:05	5.02	29
18	0.641	0.00968650	0:00:35	2.70	32
19	0.587	0.00983501	0:00:35	2.74	35
20	0.213	0.00584133	0:00:21	1.63	37
21	0.724	0.01254751	0:00:45	3.50	40
22	0.713	0.00514247	0:00:19	1.43	41
23	0.443	0.00896722	0:00:32	2.50	44
24	0.636	0.00512337	0:00:18	1.43	45
25	0.040	0.00057840	0:00:02	0.16	46
26	0.859	0.01065522	0:00:38	2.97	48
27	0.711	0.00723388	0:00:26	2.02	51
28	0.815	0.00567573	0:00:20	1.58	52
29	0.353	0.00320361	0:00:12	0.89	53
30	0.084	0.00090140	0:00:03	0.25	53
31	0.668	0.00712541	0:00:26	1.99	55
32	0.449	0.00976155	0:00:35	2.72	58
33	0.580	0.01360042	0:00:49	3.79	62
34	0.807	0.02654637	0:01:36	7.40	69
35	0.446	0.00417016	0:00:15	1.16	70
36	0.296	0.00289687	0:00:10	0.81	71
37	0.001	0.00002162	0:00:00	0.01	71
38	0.384	0.00342935	0:00:12	0.96	72

(continued)

Table 6.2 (continued)

No	Random	Log-normal formula	Arrive PERs	PROB (%)	ACUM (%)
39	0.330	0.00765315	0:00:28	2.13	74
40	0.279	0.00652828	0:00:24	1.82	76
41	0.279	0.00358722	0:00:13	1.00	77
42	0.691	0.02696885	0:01:37	7.52	85
43	0.379	0.00661347	0:00:24	1.84	86
44	0.688	0.00495803	0:00:18	1.38	88
45	0.589	0.01350218	0:00:49	3.77	92
46	0.613	0.01027003	0:00:37	2.86	94
47	0.305	0.00825954	0:00:30	2.30	97
48	0.033	0.00029565	0:00:01	0.08	97
49	0.492	0.00642460	0:00:23	1.79	99
50	0.273	0.00500267	0:00:18	1.40	100
			0:21:31	100.00	

Algorithm 2 ResultBinomial()

```
Range(“J25”).Select
ActiveCell.FormulaR1C1 ← “N”
Range(“J26”).Select
ActiveCell.FormulaR1C1 ← “1”
Range(“K25”).Select
ActiveCell.FormulaR1C1 ← “Random”
Range(“K26”).Select
ActiveCell.FormulaR1C1 ← “RAND()”
Range(“L25”).Select
ActiveCell.FormulaR1C1 ← “Binomialformula”
Range(“L26”).Select
ActiveCell.FormulaR1C1 ← “BINOM.DIST(1000 + (20 ∗ RC[−1]), 2000, 0.5, 0)”
Range(“M25”).Select
ActiveCell.FormulaR1C1 = “PROB”
Range(“M26”).Select
ActiveCell.FormulaR1C1 ← “((RC[−1]/24) ∗ 10) − 0.002”
Range(“N25”).Select
ActiveCell.FormulaR1C1 ← “RESULT”
Range(“N26”).Select
ActiveCell.FormulaR1C1 ← “IF(RC[−1] >= 0.005, ““1”", ““0”")”
Range(“O25”).Select
ActiveCell.FormulaR1C1 ← “CONT”
Range(“O26”).Select
ActiveCell.FormulaR1C1 ← “COUNTIF(RC[−1] : R[49]C[−1], 1)”
Range(“O27”).Select
```

probability distribution formula and the Montecarlo method. Later, the final results are compared with the probability obtained, if it is greater than 0.5, it is added a counter, which is counting all interactions, and finally, it shows the number of people who leave the queue for some reason.

6.2.4 Binomial Probability Distribution

It is a discrete probability distribution that counts the number of successes in a sequence of n trials of Bernoulli independent among themselves, with a fixed probability of occurrence of success between trials [23, 24].

It is applied the distribution of binomial probability to know the probability of success of finding algae within 1 quadrant. In our study, we implemented the same distribution to determine the probability of a person leaving the waiting line at the taxi station [25, 26].

$$p(x) = \binom{n}{x} p^x (1-p)^{n-x} \forall x \in \{0, 1, \ldots, n\} \tag{6.3}$$

where:

- **n**. It's the number of tests.
- **k**. It's the number of hits.
- **p**. It's the probability of success.
- **q**. It's the probability of failure.

According to the result generated by Algorithm 2. Table 6.3 shows the process of binomial distribution using random numbers to generate the probability each person has when leaving the queue and the total number of people of leaving it (Fig. 6.3).

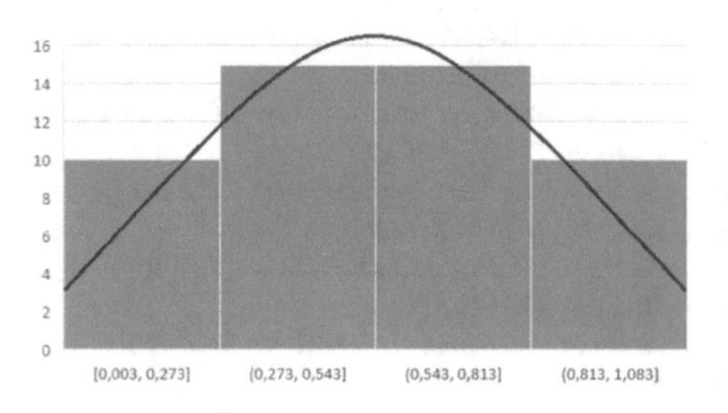

Fig. 6.3 Data from people leaving the queue

Table 6.3 Data from people leaving the queue

No	Random	Binomial formula	Prob (%)	RESULT	Contd
1	0.843	0.01381154	0.38	0	21
2	0.019	0.01783901	0.54	1	
3	0.774	0.01424622	0.39	0	
4	0.275	0.01739878	0.52	1	
5	0.281	0.01739878	0.52	1	
6	0.704	0.01466523	0.41	0	
7	0.955	0.01243539	0.32	0	
8	0.380	0.01698638	0.51	1	
9	0.142	0.01776783	0.54	1	
10	0.453	0.01645167	0.49	0	
11	0.005	0.01783901	0.54	1	
12	0.563	0.01580689	0.46	0	
13	0.142	0.01776783	0.54	1	
14	0.320	0.01720853	0.52	1	
15	0.818	0.01381154	0.38	0	
16	0.297	0.01739878	0.52	1	
17	0.580	0.01580689	0.46	0	
18	0.518	0.01614218	0.47	0	
19	0.456	0.01645167	0.49	0	
20	0.041	0.01783901	0.54	1	
21	0.954	0.01243539	0.32	0	
22	0.907	0.01290393	0.34	0	
23	0.314	0.01720853	0.52	1	
24	0.610	0.01544764	0.44	0	
25	0.111	0.01776783	0.54	1	
26	0.011	0.01783901	0.54	1	
27	0.500	0.01614218	0.47	0	
28	0.894	0.01336338	0.36	0	
29	0.370	0.01698638	0.51	1	
30	0.016	0.01783901	0.54	1	
31	0.591	0.01580689	0.46	0	
32	0.579	0.01580689	0.46	0	
33	0.583	0.01580689	0.46	0	
34	0.497	0.01645167	0.49	0	
35	0.888	0.01336338	0.36	0	
36	0.148	0.01776783	0.54	1	
37	0.641	0.01544764	0.44	0	
38	0.882	0.01336338	0.36	0	
39	0.710	0.01466523	0.41	0	

Table 6.3 (continued)

No	Random	Binomial formula	Prob (%)	RESULT	Contd
40	0.243	0.01755600	0.53	1	
41	0.376	0.01698638	0.51	1	
42	0.637	0.01544764	0.44	0	
43	0.325	0.01720853	0.52	1	
44	0.593	0.01580689	0.46	0	
45	0.416	0.01673361	0.50	0	
46	0.083	0.01782119	0.54	1	
47	0.756	0.01424622	0.39	0	
48	0.555	0.01580689	0.46	0	
49	0.683	0.01506640	0.43	0	
50	0.181	0.01767926 0.80184529	0.54 23	1	

6.3 Results

By using Algorithm 3, the results of the simulation processes are presented with the purpose of showing the effect of people arriving at the station and those who leave the queue, for different reasons, using Binomial probability distribution and log-normal probability distribution.

Tables 6.4, 6.5, and 6.6 present the total number of people entering the queue and leaving the line for different reasons marked with *.

Table 6.4 Simulation results of month 1

	Monday	Tuesday	Wednesday	Thursday	Friday	Sum	Prob
Week 1 Leave	129 21*	55 24*	129 17*	60 22*	148 15*	521 99*	0,33
Week 2 Leave	112 18*	48 20*	44 18*	46 15*	51 22*	301 93*	0,19
Week 3 Leave	82 14*	56 21*	39 20*	63 17*	59 18*	299 90*	0,19
Week 4 Leave	117 14*	92 18*	70 12*	115 24*	50 26*	444 94*	0,28
Subtotal						1565	1
Lost						376*	0,24
Total						1189	0,76

Algorithm 3 Resultad_Mes()

```
Range("B3 : I3").Select
ActiveCell.FormulaR1C1 ← "MONTH1"
Range("B5").Select
ActiveCell.FormulaR1C1 ← "WEEK1"
Range("C4").Select
ActiveCell.FormulaR1C1 ← "MONDAY"
Range("C5").Select
ActiveCell.FormulaR1C1 ← "RANDBETWEEN(30, 150)"
Range("D4").Select
ActiveCell.FormulaR1C1 ← "TUESDAY"
Range("D5").Select
ActiveCell.FormulaR1C1 ← "RANDBETWEEN(30, 150)"
Range("E4").Select
ActiveCell.FormulaR1C1 ← "WEDSNEDAY"
Range("E5").Select
ActiveCell.FormulaR1C1 ← "RANDBETWEEN(30, 150)"
Range("F4").Select
ActiveCell.FormulaR1C1 ← "THURSDAY"
Range("F5").Select
ActiveCell.FormulaR1C1 ← "RANDBETWEEN(30, 150)"
Range("G4").Select
ActiveCell.FormulaR1C1 ← "FRIDAY"
Range("G5").Select
ActiveCell.FormulaR1C1 ← "RANDBETWEEN(30, 150)"
Range("H4").Select
ActiveCell.FormulaR1C1 ← "SUM"
Range("H5").Select
ActiveCell.FormulaR1C1 ← "SUM(RC[−5] : RC[−1])"
Range("I4").Select
ActiveCell.FormulaR1C1 ← "PROB"
Range("I5").Select
ActiveCell.FormulaR1C1 ← "RC[−1]/R[8]C[−1]"
Range("B6").Select
ActiveCell.FormulaR1C1 ← "LEAVE"
Range("C6").Select
ActiveCell.FormulaR1C1 ← "Binomial(129, 0.18)"
Range("D6").Select
ActiveCell.FormulaR1C1 ← "Binomial(129, 0.18)"
Range("E6").Select
ActiveCell.FormulaR1C1 ← "Binomial(129, 0.18)"
Range("F6").Select
ActiveCell.FormulaR1C1 ← "Binomial(129, 0.18)"
Range("G6").Select
ActiveCell.FormulaR1C1 ← "Binomial(129, 0.18)"
Range("H6").Select
ActiveCell.FormulaR1C1 ← "SUM(RC[−5], RC[−4], RC[−3], RC[−2], RC[−1])"
Range("G13").Select
ActiveCell.FormulaR1C1 ← "SUBTOTAL"
Range("H13").Select
ActiveCell.FormulaR1C1 ← "R[−8]C + R[−6]C + R[−4]C + R[−2]C"
Range("I13").Select
```

$ActiveCell.FormulaR1C1 \leftarrow$ "$SUM(R[-8]C : R[-1]C)$"
$Range(\text{"}G14\text{"}).Select$
$ActiveCell.FormulaR1C1 \leftarrow$ "$LOST$"
$Range(\text{"}H14\text{"}).Select$
$ActiveCell.FormulaR1C1 \leftarrow$ "$R[-8]C + R[-6]C + R[-4]C + R[-2]C$"
$Range(\text{"}I14\text{"}).Select$
$ActiveCell.FormulaR1C1 \leftarrow$ "$RC[-1]/R[-1]C[-1]$"
$Range(\text{"}G15\text{"}).Select$
$ActiveCell.FormulaR1C1 \leftarrow$ "$total$"
$Range(\text{"}H15\text{"}).Select$
$ActiveCell.FormulaR1C1 \leftarrow$ "$R[-2]C - R[-1]C$"
$Range(\text{"}I15\text{"}).Select$
$ActiveCell.FormulaR1C1 \leftarrow$ "$R[-2]C - R[-1]C$"
$Range(\text{"}I16\text{"}).Select$

In the simulation of the first month, the total number of people entering the queue is 1,565 of which 376 leave the queue for different reasons; being 1,189 the number of people using the service, generating $1,189 gain and $376 loss.

In the simulation of the second month, the total number of people entering the queue is 1,844 of which 381 leave the queue for different reasons; being 1,463 the number of people using the service, generating $1,463 gain and $381 loss.

In the simulation of the third month, the total number of people entering the queue is 1,714 of which 382 leave the queue for different reasons; being 1,332 the number of people using the service, generating $1,332 gain and $382 loss.

Monthly the taxi station loses about 23% of the profits being equivalent to $380.

Table 6.5 Simulation results of month 2

	Monday	Tuesday	Wednesday	Thursday	Friday	Sum	Prob
Week 1	59	113	150	79	76	477	26
Leave	19	23	20	21	24	107 *	
Week 2	58	143	35	65	128	429	23
Leave	17	25	14	23	19	98 *	
Week 3	60	39	97	38	139	373	20
Leave	18	12	17	20	15	82 *	
Week 4	115	150	54	129	117	565	31
Leave	20	27	13	19	15	94 *	
Subtotal						1844	1
Lost						381 *	21
Total						1463	0,79

Table 6.6 Simulation results of month 3

	Monday	Tuesday	Wednesday	Thursday	Friday	Sum	Prob
Week 1	32	44	101	52	94	323	19
Leave	14	19	22	24	20	99 *	
Week 2	138	90	145	123	59	555	0,32
Leave	16	22	20	15	17	90 *	
Week 3	64	114	72	69	140	459	27
Leave	21	14	17	20	18	90 *	
Week 4	100	50	65	33	129	377	22
Leave	22	24	19	17	21	103 *	
Subtotal						1714	1
Lost						382 *	22
Total						1332	0,78

6.4 Conclusion

Analyzing the results obtained in 3 months, we can conclude that the taxi cooperative can lose about 23% of the profits in each month If you keep the same way of managing. It can be solved in different ways, for example, the acquisition of new vehicles in order to meet daily demand, preventing people from leaving the queue.

Finally, it can be concluded that using the log-normal probability distribution tools and the Binomial probability distribution, Montecarlo and Visual Basic for Excel, the simulated data of the consecutive three months can be obtained and their possible losses. Thus, it is recommended for future work a meticulous analysis of one year to make the results more accurate.

References

1. Torres, Alvaro, and Carolina Tranchita. 2004. ¿ inferencia y razonamiento probabilístico o difuso? *Revista de Ingeniería* 19: 158–166.
2. Brendel, William, Alan Fern, and Sinisa Todorovic. 2011. Probabilistic event logic for interval-based event recognition. In *2011 IEEE Conference on Computer Vision and Pattern Recognition (CVPR)*, 3329–3336. IEEE.
3. Pearl, Judea. 2014. *Probabilistic reasoning in intelligent systems: networks of plausible inference*. Elsevier.
4. Schum, David A. 2001. *The evidential foundations of probabilistic reasoning*. Northwestern University Press.
5. Pearl, Judea. 1987. Evidential reasoning using stochastic simulation of causal models. *Artificial Intelligence* 32 (2): 245–257.
6. Tarifa Enrique Eduardo. 2001. *Teoría de modelos y simulación*. Facultad de Ingeniería: Universidad de Jujuy.
7. Zapata, Carlos J. 2010. Análisis probabilístico y simulación. *Grupo de investigación en planeamiento de sistemas eléctricos*. Universidad Tecnológica de Pereira. Pereira–Colombia.

8. Badii, M.H., and J. Castillo. Distribuciones probabilísticas de uso común. *Revista Daena (International Journal of Good Conscience)* 4 (1).
9. Martínez, Alexei Gómez Eguiarte, and Gabriel de las Nieves Sánchez Guerrero. Aplicación de funciones de distribución continuas para modelar la demanda de pasajeros en una línea de tren ligero. *Contaduría y administración* 61 (1): 159–175.
10. Lojano, Juan Pablo, Álex Rojas, and Vilma Rojas. 2017. Un modelo híbrido de probabilidad de elección para la estimación de la demanda de quitocable. *Maskana* 8: 219–228.
11. Orozco Juan Sebastián, and Fernando Antonio Arenas. 2013. Aproximación al desarrollo de un sistema de transporte masivo a través de la dinámica de sistemas. *Sistemas and Telemática* 11 (24): 91–106.
12. Corral de Franco,Yadira Josefina. 2009. Validez y confiabilidad de los instrumentos de investigación para la recolección de datos.
13. Mielnisuk, Nicolás O, Sonia I Mariño, and Romina Yolanda Alderete. 2016. Diseño de un entorno colaborativo: Una aplicación para apoyar el aprendizaje de técnicas de modelado y simulación de la teoría de colas. *IE Comunicaciones: Revista Iberoamericana de Informática Educativa* (23): 2.
14. Mielnizuk, Nicolás O, Sonia I. Mariño, and Romina Y. Alderete. 2016. Simuladores para apoyar el aprendizaje de la teoría de colas. *Revista Premisa* 18 (69).
15. Schnarwiler, Jorge L., Justo J. Roberts, Pedro O. Prado, Silvia L. Bocero, and Agnelo M Cassula. Centrales solares fotovoltaicas en áreas con pasivos ambientales.
16. Chi, Rosa Imelda Garcia., Arturo Eguia Alvarez, Gloria Emilia Izaguirre Cardenas, et al. 2015. Uso de la herramienta de software promodel como estrategia didáctica en el aprendizaje basado en competencias de simulación de procesos y servicios. *TECTZAPIC* (1).
17. Baran, Sándor, and Sebastian Lerch. 2015. Log-normal distribution based ensemble model output statistics models for probabilistic wind-speed forecasting. *Quarterly Journal of the Royal Meteorological Society* 141 (691): 2289–2299.
18. Khandelwal, Vineet, et al. 2014. A new approximation for average symbol error probability over log-normal channels. *IEEE Wireless Communications Letters* 3 (1): 58–61.
19. Martínez, Iveth, and Mihaela Stegaru. 2010. Análisis preliminar del servicio del transporte público de el corregimiento de ancon área revertida (2005–2007). 12 (1): 93–102.
20. Schwartz, Stuart C. and Yu-Shuan Yeh. 1982. On the distribution function and moments of power sums with log-normal components. *Bell System Technical Journal* 61 (7): 1441–1462.
21. Gulisashvili, Archil, Peter Tankov, et al. 2016. Tail behavior of sums and differences of log-normal random variables. *Bernoulli* 22 (1): 444–493.
22. Muñoz, David F., César Ruiz, and Sven Guzman. 2016. Comparación empírica de varios métodos para estimar los parámetros de la distribución lognormal con traslado. *Información tecnológica* 27 (3): 131–140.
23. Reyes, Hortensia, and F. Almendra. 2015. Problemas al usar la aproximación normal n intervalos de confianza suponiendo datos bernoulli. XXV Simposio Internacional de Estadística.
24. Zamora, Larisa, and Jorge Díaz. 2019. Empleo del paquete exprep para repetición de ensayos de bernoulli en la enseñanza de las probabilidades. *Revista Digital: Matemática, Educación e Internet* 19 (1).
25. Alvarado, Hugo, and María Lidia Retamal. 2014. Representaciones de la distribución de probabilidad binomial.
26. García, Javier, Ernesto A. Sánchez, and M. Mercado. 2017. Razonamiento probabilístico de estudiantes de bachillerato frente a una situación binomial.

Chapter 7
Case Study: Project-Based Learning to Evaluate Probability Distributions in Medical Area

This study presents the use of probability distributions and the theory of systems simulation applied to real-life problems, with the aim of giving researchers and Students a guide to facilitate its application within of investigative work. It was carried out project-based learning (PBL) through a project in the classroom in order to help students to recognize, develop and apply feasibly the different types of probability distributions in real life problems.

7.1 Introduction

Project-Based Learning (PBL) is a methodology of teaching and learning focused on tasks, and it is essential to be clear that this method promotes individual and autonomous learning within a work plan defined by objectives and procedures. To achieve this goal, students are responsible for their learning,in other words, discover their preferences and strategies throughout the process [1–5].

In order to have the PBL methodology clearer, it has been considered its application in the field of probability theory, in such a way that the student achieves to deepen in the learning of the distributions of probability, and can obtain the necessary significant knowledge that will help him to solve problems of real situations related in your professional field [6–9].

It was evident the students in the class of systems simulation at the University of Guayaquil do not have a solid knowledge of probability issues. In other words, the students do not have sufficient experience to simulate probability distributions by using methods such as inverse transformed and Montecarlo simulation, since they do not have the adequate fluency for the resolution and application of these issues in real-life situations [10–13].

L. Cevallos-Torres and M. Botto-Tobar, *Problem-Based Learning: A Didactic Strategy in the Teaching of System Simulation*, Studies in Computational Intelligence 824, https://doi.org/10.1007/978-3-030-13393-1_7

On the other hand, it is essential that the teacher enable continuous feedback and evaluation. In other words, he/she must create an optimal learning environment guiding the process, encouraging the use of meta cognitive strategies and reinforcing the efforts both individual as a group, keeping a thorough follow-up the design of projects. Furthermore, the professor has to give feedback on the contents and to carry out a group level evaluation of the learning acquired by each student [14–17].

7.1.1 Related Work

Osorio et al. in [18] proposes the use of different techniques and programs for the learning of probability, as well as the concepts related to it. In this situation, learning is not enough because students do not have a deeper understanding of the theory of odds. Therefore, they do not develop skills to simulates events from real life [19–21].

Burbano et al. in [22] the main problem is that the simulations were performed directly by using computing programs, and the students are not able to do the process on their own [23–25].

Flowers-Cano et al. in [26] used the distributions Gamma, Weibull, Gumbel, Log-normal and Log-logistic jointly with the simulated data simulates, to see if these are adapted to the probability distribution and in this way reject or accept the simulation by using evidence such as the Anderson-Darling [27, 28].

7.2 Case Study

For the development of teaching process it was considered that the sixth-semester students of the simulation course had a cautious attitude regarding the use of probability distributions; due to many associated factors, such is the case of information provided by colleagues who already approved the subject and are in the highest semesters.

It is essential to be clear that the student feels uncertainty and even a little fear of subject because they have not applied appropriate pedagogical strategies to help him develop his cognitive skills that lead to critical thinking. The current model that the student receives takes him to a by the rote process and merely the use of a guiding text, which does not allow the development of investigative capacities by not knowing the positions of other authors regarding the subjects treated. This situation allowed us to reflect on the pedagogical model, the methodologies and teaching strategies used in our teaching [29, 30].

For work on project-based learning projects to the application of probability distributions in simulation, it is proposed to solve a problem in the medicine area. A sample of patients arriving at a particular office was collected through the observation in an eight-hour period in two working days. It was also considered to determine the period that each patient took to arrive at the clinic as well as the type of disease that the patient has.

7.2.1 *Evaluation*

To properly carry out the methodology PBL, it is important to have clear the following:

1. To analyze the comprehension of topics, the participation of the students should be observed during the development of the activity.
2. To ask students what was the most challenging question and why? It will allow visualizing the handling of the concepts.
3. It is important to observe and analyze the different processes applied by the students.
4. The acceptance of the activity (positive and negative aspects) should be analyzed.
5. To complement the activity, it is essential to take into account the recommendations and suggestions on the part of the students for the next opportunity.

7.2.2 *Data Collection*

The PBL methodology looks for the students to take an active role. This method motivates them to learn more, integrating probability distribution and simulation knowledge in any area. It aims for the students to leave that passive role, in which they received information, memorized it and over time was forgotten.

7.2.3 *Activity Process*

This stage applies to a real problem, where the students begin to collect the patients' arrival information in a doctor's office, as well as the type of disease they suffer. We obtained a small sample to the application of random numbers and Montecarlo simulation.

To simulate data from a real sample, you must determine the behavior that follows this data. For this, it should know which kind of probability distribution is the most suited; Then, we used the inverse transform, to be able to determine a formula that generates as many values as possible. Also, the distribution of Weibull and the empirical probability distribution was used.

7.2.3.1 Probability Distribution of Weibull

It is a continuous probability distribution that allows modeling fault events (or other event) in systems when they are proportional in a time [31, 32].
Distribution function applied:

$$f(x) = \infty B^{-\infty} t^{\infty-1} e^{-\left(\frac{t}{\beta}\right)^{\infty}} \tag{7.1}$$

Accumulated distribution:

$$F(x) = 1 - e^{-\left(\frac{x}{B}\right)^{\infty}} \tag{7.2}$$

7.2.3.2 Inverse Transform Application

Since the inverse transformed can only be performed with continuous probability distributions that have a cumulative function; the probability distribution of Weibull is the most suitable for this research. Its process is as follows:

$$\begin{gathered}
F(x) = 1 - e^{-\left(\frac{x}{B}\right)^{\infty}} \\
r = 1 - e^{-\left(\frac{x}{B}\right)} \\
1 - r = e^{-\left(\frac{x}{B}\right)} \\
\log(1 - r) = \log\left(e^{-\left(\frac{x}{B}\right)}\right) \\
\log(1 - r) = -\left(\frac{x}{B}\right)\log(e) \\
\log(1 - r) = -\left(\frac{x}{B}\right) \\
-B(\log(1 - r)) = x \\
x = -B(\log(1 - r)) \\
x = -B * \log(r)
\end{gathered}$$

where:
B = The average time of arrival of patients to the particular office.
r = Random number.
x = Random number applicable to the probability function. In this case, these numbers are the arrival times between each patient.

7.2.3.3 Empirical Probability Distribution

Its equation is as follows:

$$f(x) = \frac{\#(x_i \epsilon A)}{n} \tag{7.3}$$

It's a probability distribution discrete. It models the randomness of a variable that can only take specific values. This formula was used because it explains in a simple way that it is going to divide the number of appearances of a x_i. In our case, the disease which belongs to A set (our disease set) to n (our total number of cases) [33].

7.2.3.4 Other Applications of This Probability Distribution

In studies as cardiovascular disease, risk factors in the area of health [34]. An empirical distribution is used to analyze the results of the clinical histories and surveys carried out.
Distribution function applied:

$$f(x) = \begin{cases} 4\%, & \text{Intestinal infection} \\ 42\%, & \text{Seasonal influenza} \\ 52\%, & \text{Other diseases} \end{cases} \tag{7.4}$$

Accumulated distribution:

$$F(x) = \begin{cases} \text{Intestinal infection}, & \textit{If } 0 \le X < 0.0047 \\ \text{Seasonal influenza}, & \textit{If } 0.0047 \le X \le 0.428 \\ \text{Other diseases}, & \textit{If } 0.428 \le X \le 1 \\ 0, & \text{Otherwise} \end{cases} \tag{7.5}$$

where:
X = variable to be replaced by a random number to get a random disease.
Other diseases = Rare or less common diseases than intestinal infection and seasonal influenza.

By this, it was obtained the followings results (Table 7.1):

From this data, the average is calculated which gets a value of 62.19. It will be the value to use in the formula to simulate a more significant amount of data to obtain a better precision in the results.

The values to determine the empirical probability appearance of diseases are presented in Table 7.2.

For other illnesses, we refer to rare or unusual diseases that do not represent a significant sample by themselves to make their accurate simulation.

7.3 Results and Discussion

With the design of these activities for the learning of these probability distributions through the PBL methodology, it is intended that the teacher and the student find a complementary approach to acquire the mathematical knowledge of these distributions so that they might be used more efficiently in the learning of the subject of simulation of systems.

It is worth to remember that we took a few data. However, it is necessary to proceed to obtain a more significant amount of data, that is why simulation is made based on the data collected previously. For this, we use, the inversed transformation to determine a formula that generates as many values as possible. It was done by

Table 7.1 Patient arrival time to a particular office

Day 1			Day2		
Patient	Time	Disease	Patient	Time	Disease
1	6	Other diseases	1	19	Seasonal influenza
2	14	Other diseases	2	75	Seasonal influenza
3	87	Seasonal influenza	3	18	Seasonal influenza
4	165	Seasonal influenza	4	52	Other diseases
5	14	Seasonal influenza	5	184	Other diseases
6	56	Infection Intestinal	6	17	Other diseases
7	161	Other diseases	7	66	Other diseases
8	28	Seasonal influenza	8	26	Other diseases
9	26	Other diseases	9	22	Other diseases
10	100	Seasonal influenza	10	21	Other diseases
			11	149	Seasonal influenza

Table 7.2 Frequencies' table for the different diseases in the sample

Case	Frequency	Relative frequency (FR)	FR accumulated
Intestinal infection	1	0.047619048	0.047619048
Seasonal influenza	9	0.428571429	0.476190476
Other diseases	11	0.523809524	1
Total	21	1	

Table 7.3 Simulation of a working day

# Patient	Time hold on	Type disease
1	253	Another
2	33	Intestinal infection
3	56	Seasonal influenza
4	28	Another
5	7	Another
6	36	Seasonal influenza
7	123	Seasonal influenza
8	8	Another
9	75	Another

using a distribution of Weibull, which was the probability distribution used in the arrival time between patients. Besides, we simulate waiting times and the possible illnesses that each patient contains through the empirical distribution accumulated.

By simulating data based on a probability distribution, we ensure that the results obtained through simulation will be similar to the input data, previously obtained in the field, this is something that is wanted to do a simulation because it tells us that We are doing the process correctly.

Table 7.4 Results of the simulation of a working day

Day 1		
Patient	Time	Disease
1	48	Seasonal influenza
2	2	Other diseases
3	24	Other diseases
4	20	Seasonal influenza
5	11	Intestinal infection
6	24	Seasonal influenza
7	132	Other diseases
8	123	Seasonal influenza
9	96	Other diseases
10	34	Other diseases
11	55	Other diseases
12	23	Other diseases
13	37	Other diseases
14	9	Other diseases

Algorithm 1 simulateDay()

```
DeleteDay
band ← True
Dn1 ← 3
Dn2 ← 9
sum ← 0
while band = True do
  val ← Sheet2.Cells(Rnd * 100 + 1, Rnd * 100 + 2)
  val2 ← Sheet3.Cells(Rnd * 100 + 1, Rnd * 100 + 2)
  if val > 0 then
    sum ← sum + val
    if sum ≥ 660 ORDn1 ≥ 22 then
      band ← False
    else
      loadData(Dn1, Dn2, val, val2, 2)
      Dn1 ← Dn1 + 1
    end if
  end if
end while
```

By observing Table 7.3, it indicates that nine patients will be cared for, of which three will have seasonal influenza (common influenza), one will have an intestinal infection and the others come with other types of illnesses. This is normal and easy to see day to day result of any medical establishment. Several conclusions can be obtained from such a result, and for example, in this case, we could say that there is no need to make any changes. However, we do not recommend concluding such a small simulation. Table 7.4 shows the obtained values from simulation. Algorithm 1 allows to generate the simulation of a working day.

Table 7.5 Average time in 25 one-month simulations

Month	Average
1	69.8685
2	78.1293
3	58.5625
4	65.2532
5	65.0144
6	62.3604
7	67.2957
8	64.8036
9	62.2071
10	64.781
11	64.3979
12	61.5663
13	64.549
14	56.5884
15	62.4214
16	55.3251
17	64.6355
18	64.7992
19	72.6303
20	57.139
21	55.2761
22	60.58
23	65.7796
24	57.3675
25	66.2452
Total mean	63.503

Before to start analyzing and draw conclusions from the simulated results; we must make sure that these are correct. There are several ways to validate whether or not you are simulating the system correctly, but because of its easy implementation, the mean will be checked.

To validate the system is giving the correct values, it must find the time average (μ) in each simulated day and verify if the obtained value is not far from the average of our original data. Besides, if we want to confirm that our system is valid, we could considerate the following:

1. The average of simulated data will hardly become the same as the average of original data since these data are not the same, what should be verified is that the variability of those data is not far from the average.
2. The mean, being an average of all values can be hugely affected by atypical values, in some cases it will give us values far from the average of our original data, this

Table 7.6 20-week simulation results

Index	People served	People who arrived with seasonal influenza	People who arrived with intestinal infection	People who arrived other illnesses
1	62	12	6	32
2	59	14	2	34
3	64	21	4	29
4	68	22	5	34
5	64	20	3	24
6	63	27	1	26
7	65	26	5	22
8	72	19	5	34
9	69	21	3	31
10	74	17	4	30
11	72	19	4	34
12	71	27	3	30
13	80	26	4	37
14	78	15	2	32
15	75	18	4	32
16	71	26	3	24
17	71	26	5	23
18	66	14	4	25
19	68	11	3	24
20	76	23	3	38

does not mean that the model is wrong unless something is repeated too many in time.

3. It must have the mean of a significant sample of simulations because if we have very little information can be misinterpreted data

Given this, we proceed to validate that the system is getting correct data:

$$\text{Average of original data} = \mu = 62.35$$

Table 7.5 shows the daily mean of the patients' arrival time in 25 different simulations in one month. (The simulation consisted in simulating their days.)

The averages total (Table 7.5) of simulations is similar to the average of original data collected in the field, so we can say that the simulation if it is giving values that represent or resemble the reality.

Having assured us that correct results are being obtained, we then proceed to use these results to make decisions in the real world.

Algorithm 2 simulateMontLab()

```
DeleteMonth
Mmonth ← 1
while Mmonth ← 4 do
  M week ← 58 + (24 ∗ (Mmonth − 1))
  Sday ← 1
  Dn2 ← 9
  while Sday ≤ 5 do
    band ← True
    Dn2 ← M week
    sum ← 0
    while band = True do
      val ← Sheet2.Cells(Rnd ∗ 100 + 1, Rnd ∗ 100 + 2)
      val2 ← Sheet3.Cells(Rnd ∗ 100 + 1, Rnd ∗ 100 + 2)
      if val > 0 then
        sum ← sum + val
        if sum ≥ 660 ORDn1 ≥ M week + 19 then
          band ← False
        else
          loadData(Dn1, Dn2, val, val2, M week − 1)
          Dn1 ← Dn1 + 1
        end if
      end if
      Dn2 ← Dn2 + 3
      Sday ← Sday + 1
    end while
    Mmonth ← Mmonth + 1
  end while
end while
```

To facilitate decision making and to obtain results with better precision, many simulations were carried out; this was done thanks to the implementation and development of a small algorithm, performed in Visual Basic programming for Excel, whose Result presents us the number of patients attended for a month.

With the obtained data from 20-week simulations, it is already possible to make decisions, for example, to know if we should focus on the needs of a group of patients with a particular disease, and thus be able to prepare for possible epidemics. Table 7.6 shows simulations 7, 16 and 17 that the disease that most affected the population was seasonal influenza. It means that in the real-life case, there is a high degree of possibility that in 20-weeks appear small flu epidemics.

7.4 Conclusion

The methodology PBL was well accepted by the students in the engineering degree of computer systems. The students agreed to change the role in the classroom to be more active and being able to develop skills in the subject of system simulation.

On the other hand, it was possible to determine that the learning of probability distributions applied to the subject of system simulation is essential and useful to use it in issues of real-life.

References

1. Savery, John R. 2015. Overview of problem-based learning: Definitions and distinctions. *Essential Readings in Problem-Based Learning: Exploring and Extending the Legacy of Howard S. Barrows* 9: 5–15.
2. Storer, Terry. 2018. The effect of project based learning on the creativity of elementary students. Ph.D. thesis, Wilkes University.
3. Bilgin, Ibrahim, Yunus Karakuyu, and Yusuf Ay. 2015. The effects of project based learning on undergraduate students' achievement and self-efficacy beliefs towards science teaching. *Eurasia Journal of Mathematics, Science and Technology Education* 11 (3): 469–477.
4. Lee, Jean S., Sue Blackwell, Jennifer Drake, and Kathryn A. Moran. 2014. Taking a leap of faith: Redefining teaching and learning in higher education through project-based learning. *Interdisciplinary Journal of Problem-Based Learning* 8 (2): 2.
5. Egilmez, Gokhan, Dusan Sormaz, and Ridvan Gedik. 2018. A project-based learning approach in teaching simulation to undergraduate and graduate students.
6. Moreno, Amable, José Cardeñoso, and Francisco González-García. 2015. Los significados de la probabilidad en los profesores de matemática en formación: un análisis desde la teoría de los modelos mentales.
7. Estruch, Vicente D., Francisco J. Boigues, and Anna Vidal. 2017. Un recorrido de estudio e investigación para el aprendizaje del concepto devariable aleatoria discreta mediante métodos de monte carlo. *Modelling in Science Education and Learning*, vol. 10, 67–84. Universitat Politècnica de València.
8. García, José Miguel Contreras, María del Carmen Batanero Bernabeu, María del Mar López Martín, and Magdalena S. Carretero Rivas. 2015. Los problemas de probabilidad propuestos en las pruebas de acceso a la universidad en andalucía. *Areté: Revista Digital del Doctorado en Educación de la Universidad Central de Venezuela* 1 (1): 39–60.
9. Jean, Neal, Michael Xie, and Stefano Ermon. 2016. Semi-supervised deep kernel learning. In *NIPS Bayesian deep learning workshop.*
10. Calvo, Lourdes, and Cristina Prieto. 2016. The teaching of enhanced distillation processes using a commercial simulator and a project-based learning approach. *Education for Chemical Engineers* 17: 65–74.
11. Seman, Laio Oriel, Romeu Hausmann, and Eduardo Augusto Bezerra. 2018. Agent-based simulation of learning dissemination in a project-based learning context considering the human aspects. *IEEE Transactions on Education*, 61 (2): 101–108.
12. Calderón, Alejandro, and Mercedes Ruiz. 2014. Evaluación automática en dirección y gestión de proyectos software a través de un juego basado en simulación. *Jornadas de Enseñanza Universitaria de la Informática (20es: 2014: Oviedo).*
13. Huerta Palau, M. 2015. La manera de resolver problemas de probabilidad por simulación. *Didáctica de la Estadística, Probabilidad y Combinatoria*, 53.
14. Juan, Elena Sánchez, Peregrina Del Carmen Coll Aliaga, Damián Ginestar Peiro, and Esther Sanabria Codesal. 2017. Transformada de laplace con mathematica.
15. Ruiz Moreno, Leonsio, Patricia Camarena Gallardo, and Socorro del Rivero Jiménez. 2016. Prerrequisitos deficientes con software matemático en conceptos nuevos: transformada de laplace. *Revista mexicana de investigación educativa* 21 (69): 349–383.
16. Martínez-Marín, Francisco Alejandro, and Irma Adriana Cantú-Munguía. 2017. Manejo de la simulación en la enseñanza de la ingeniería. *Revista Educación en Ingeniería* 12 (24): 58–62.

17. Gutiérrez Rodas, Daniel Alejandro. 2016. Diseño e implementación de una herramienta de software para el análisis de confiabilidad de sistemas eléctricos de potencia basado en el método de simulación de montecarlo. B.S. thesis, Quito.
18. Osorio Angarita, María Alejandra, Augusto Suárez Parra, and Carmen Constanza Uribe Sandoval. 2013. Revisión de alternativas propuestas para mejorar el aprendizaje de la probabilidad. *Revista Virtual Universidad Católica del Norte* 1 (38): 127–142.
19. Vélez, Lizzette Mestey. 2015. La simulación clínica y su relación con el desarrollo de pensamiento crítico y la toma de decisiones. *Revista Experiencia Docente* 2 (2): 37–42.
20. Ruiz-Gómez, José Luis, José Ignacio Martín-Parra, Mónica González-Noriega, Carlos Godofredo Redondo-Figuero, and José Carlos Manuel-Palazuelos. 2018. La simulación como modelo de enseñanza en cirugía. *Cirugía Española* 96 (1): 12–17.
21. Estelrrich, Pedro Martín Alfredo. 2017. Enseñanza de la Otorrinolaringología en el curriculum de la carrera de Medicina de la Facultad de Ciencias Médicas de la Universidad Nacional de La Plata, utilizando la simulación clínica y el aprendizaje basado en la resolución de problemas. Ph.D. thesis, Facultad de Ciencias Médicas.
22. Pantoja, Víctor Miguel Ángel Burbano, Jesús Enrique Pinto Sosa, and Margoth Adriana Valdivieso Miranda. 2015. Formas de usar la simulación como un recurso didáctico. *Revista Virtual Universidad Católica del Norte* (45): 17–37.
23. Velasco, Juan, and Laura Buteler. 2017. Simulaciones computacionales en la enseñanza de la física: una revisión crítica de los últimos años. *Enseñanza de las Ciencias* 35 (2): 0161–0178.
24. Nuñez, J., J. Cepeda, and G. Salazar. 2015. Comparación técnica entre los programas de simulación de sistemas de potencia digsilent powerfactory y pss/e. *Revista Técnica Energía* (11).
25. López Fernández, Ana Gloria, Jaime Cruañas Sospedra, Adys H. Salgado Friol, Lourdes H. Lastayo Bourbón, and Carlos Manuel Pérez Yero. 2015. La enseñanza de la estadística utilizando herramientas dinámicas computacionales. *Revista Habanera de Ciencias Médicas* 14 (2): 218–226.
26. Flowers-Cano, Roberto S., Robert Jeffrey Flowers, and Fabián Rivera-Trejo. 2014. Evaluación de criterios de selección de modelos probabilísticos: validación con series de valores máximos simulados. *Tecnología y ciencias del agua* 5 (5): 189–197.
27. Hernández, Giovany Orozco, Jhon Jairo Olaya Flórez, and Elisabeth Restrepo Parra. 2014. Fundamentos de simulación de materiales por medio del método de monte carlo. *Avances: Investigación en Ingeniería* 11 (1): 100–110.
28. García Martínez, Tomás, José Luis Fernández Sánchez, and José Mira Mcwilliams. 2018. Simulación de monte carlo de los riesgos de proyecto para el desarrollo de software de domótica.
29. Peralta, Wilian M. 2015. El docente frente a las estrategias de enseñanza aprendizaje. *Revista Vinculando*.
30. Merino, José M., María E. Mathiesen, Olga Mora, Ginette Castro, and Gracia Navarro. 2014. Efectos del programa talentos en el desarrollo cognitivo y socioemocional de sus alumnos. *Estudios pedagógicos (Valdivia)* 40 (1): 197–214.
31. Bourguignon, Marcelo, Rodrigo B. Silva, and Gauss M. Cordeiro. 2014. The Weibull-G family of probability distributions. *Journal of Data Science* 12 (1): 53–68.
32. Ojeda, Cesar, and Luz Adriana Pereira. 2018. Comparación de intervalos de confianza obtenidos mediante aproximación normal y bootstrap para los parámetros de la distribución weibull. *Comunicaciones en Estadística* 11 (1): 63–75.
33. Chen, Chunlin, Robert C. Elliott, and Witold A. Krzymień. 2018. Empirical distribution of nearest-transmitter distance in wireless networks modeled by matérn hard core point processes. *IEEE Transactions on Vehicular Technology* 67 (2): 1740–1749.
34. Pérez Iglesias, Silma, Godofredo Maurenza González, Luis Nafeh Abi-Resk, and Víctor M. Romero González. 1998. Enfermedad cerebrovascular: factores de riesgo en un área de salud. *Revista Cubana de Medicina General Integral* 14 (2): 135–140.

Chapter 8
Case Study: Probabilistic Estimates in the Application of Inventory Models for Perishable Products in SMEs

The goal of this study is to create an inventory management model that will be able to estimate the control of the perishable products of a business by using probabilistic distributions. The problem arises since the stores or mini markets owners have not defined a clear concept in how to maintain an inventory in optimal conditions, especially regarding perishable products because they only have a maximum time of a week to be sold them. To solve this problem, we used specific algorithms that will help us in the handling of large amounts of data such as Monte Carlo simulation, so that we were able to use probabilistic distributions to determine the economic order quantity (EOQ) of perishable products based on weekly demand. As a result, we obtained an inventory management model, which is based on the maximum and minimum quantity of products to be ordered by the company, and also a model EOQ with an adjustment in the reorder point which it was verified a small increment in business sales by 5% during the first 11 days.

8.1 Introduction

Any company, either production, marketing or services requires the products supply of products to carry out its production and/or sale activities. Therefore, inventory management constitutes one complex logistical aspect, and this study aims to draw up a guide for the implementation of processes regarding inventory management in products of rapid deterioration.

Quantitative techniques have supported the typical decisions to be taken in respect of the inventory, in this case, in particular, probability distributions allow to obtain an estimate of possible outcomes. In other words, to help to estimate or prevent events, whether they applied in any field of research. For this study, the field of has been considered inventories of perishable products, because the inadequate use, would generate certain difficulties for the managers of a business, because they do not have an adequate control of this type of inventory based on supply and demand, also

L. Cevallos-Torres and M. Botto-Tobar, *Problem-Based Learning: A Didactic Strategy in the Teaching of System Simulation*, Studies in Computational Intelligence 824, https://doi.org/10.1007/978-3-030-13393-1_8

because they do not have much time to be able to sell them, so they decide to buy them in less quantity because of the fear of being damaged.

The important thing of this work is to make known about techniques to implement an inventory management model based on the daily or weekly demand of these products, as well as determine the optimal quantity of the order that will be made, determine the point of optimal reorder that will not generate shortages in the business.

8.1.1 Related Work

Causado Rodriguez et al. in [1] developed a proposal to improve the inventory system for a food trader in a city in Colombia. To achieve the reduction of inventory costs and have an increase in the economic benefit of the organization, through the planning and control of purchases and sales of products. The process applied by [1] consisted of classification of products by the ABC model according to the importance of each product in the sales total of the distributor. Later, the authors used the model of economic order quantity (EOQ), to systematize the periodic counts in the stored products. Also, to determine the optimal amount of orders and the exact moment in which merchandise could be ordered from suppliers and the minimum quantities of reordering, however, the author's did not use computer tools or simulation to help to generate more values to measure the proposed model.

Escobar et al. in [2] raised a policy with security stock to a probabilistic model that maximizes the expected maximum utility, considering that the products are perishable, and therefore, it might be stored for a maximum number of days. They proposed a methodology based on the Monte Carlo simulation with computational experiments using real instances obtained from a fish marketing company in the Colombian market. They showed the efficiency and effectiveness of their methodology based on expected net utility maximization.

Jara et al. in [3], the authors showed the application of a method to calculate the economic order quantity (EOQ), and the reorder point (ROP) for an international trading company of auto parts. The goal was to reduce back orders and improve customer service. As a result, they managed to adjust the logistic costs by having higher sales, and therefore, a more significant economic benefit. Although the company had an inventory system with EOQ and ROP, this was not very effective; therefore, they had to recalculate and update the inventory control system with the new values of EOQ/ROP and with each of the products of the company.

Rodriguez et al. in [4] presented in their study the application of business-oriented computational technologies with adjustments of mathematical models for Inventory Control. They evaluated the performance of such techniques that considerably helped to visualize the movement of each product within the warehouse, and in turn with those data they were able to determine the optimal quantity to order, the frequency of the orders, the period of each order and the products with the highest incidence using the ABC classification model.

References [2–4] do not use complex algorithms to handle or simulate large amounts of data. Consequently, they based on the mathematical part. For this reason, we propose in this study, the use of simulation models and computer tools, to simulate the sales of products (bananas) for a week, and know what decisions can be taken based on the results obtained.

8.2 Case Study

The owners of stores or mini markets do not have defined a concept that allows them to obtain the necessary quantity of perishable products for their establishment, without that they reach a state of rot. Therefore, generate economic losses in the store, to obtain a better knowledge of the process that is being carried out, it has been considered a study In the Inventory Systems Management [5, 6].

The objective of this study was to determine how many people enter a perishable products store to make a purchase. The considered variable was the time of arrival of the person to the premises in two hours, and another variable considered was the purchase of the product. In other words, whether the person upon entering the establishment wearing one of the products perishable, in this case (banana).

8.2.1 Inventory Management

The management of inventories constitutes one of the most complex functions of the organizations since it implies to keep stored products to protect against uncertainties at the lowest possible cost and to satisfy a demand in the future [1, 7, 8].

In mathematical terms, we can say inventory management raises from maximizing profitability and minimizing costs. Thus, the inventory management is defined as a set of decisions, rules, guidelines and/or policies through which the inventory levels to be maintained are determined when stocks are to be replenished and the size by which orders are to be made [9–12].

8.2.2 EOQ Inventory Model

It is the fundamental model for inventory control. Inventory in the process and finished product constitute an aspect of great importance for the organization and are a starting point for strategic decisions of the company. In this sense, inventory management becomes a tool to register the amounts owned by the company; which play a fundamental role in the stage of supply, and the development of demand, resulting in reliable States in the control of materials and products [13–15].

Parameters:
$Q^* =$ Optimal quantity of orders.
$D =$ Demand.
$S =$ The cost of issuing an order/cost for ordering.
$H =$ The cost associated with maintaining a unit in inventory in one year.
Formula (8.1) will allow determining the optimal quantity of orders to be made in order not to have a loss of products [16, 17].

$$ROP = d * LT \tag{8.1}$$

Parameters:
$ROP =$ Reorder Point.
$d =$ Daily demand.
$LT =$ Wait time.
To get the reorder point, we must have the daily demand that will be multiplied by the timeout.

$$d = \frac{D}{365} \tag{8.2}$$

where:
$d =$ Daily demand.
$D =$ Demand.
To obtain the daily demand of the product, we must divide the product demand for 365.

$$t = \frac{Q}{D} * \#working\ days. \tag{8.3}$$

We have:
$t =$ Cycle time.
$Q =$ Optimal quantity of orders.
$D =$ Demand.
To obtain the cycle time, we must have the optimal quantity of orders that will be divided for the demand and then multiplied by the number of working days.

$$CTA = H\left(\frac{Q}{2}\right) + \left(\frac{D}{Q}\right) \tag{8.4}$$

We have:
$t =$ Cycle time.
$Q =$ Optimal quantity of orders.
$D =$ Demand.
To obtain the cycle time, we must have the optimal quantity of orders that will be divided for the demand and then multiplied by the number of working days.

On the other hand, it is essential to be clear that data which were taken as part of the sample, and the management of the inventory was performed during a five

Algorithm 1 Use of the EOQ model algorithm that will help optimize the loss of perishable products from the store or mini-market to which is conducting the study.

```
for i = 120 To 130 do
  demand ← Cells(112, k).Value
  if z = 1 then
    if d = p then
      Cells(i, 2).Value ← invInitial + Cells(114, 6).Value
      Cells(i, 2).Select
    else
      Cells(i, 2).Value ← invInitial
      Cells(i, 2).Select
    end if
  else {z = 0}
    Cells(i, 2).Value ← invInitial
    Cells(i, 2).Select
  end if
  Cells(i, 3).Value ← demand
  Cells(i, 3).Select
  invFinal ← Cells(i, 2).Value − Cells(i, 3).Value
  Cells(i, 4).Value ← invFinal
  Cells(i, 4).Select
  if Cells(i, 4).Value < Cells(114, 6).Value then
    Cells(i, 5).Value ← 1
    Cells(i, 5).Select
  else
    Cells(i, 5).Value ← 0
    Cells(i, 5).Select
  end if
end for
```

day period. The sample taken was not sufficient to obtain a favorable outcome, so we proceeded to enter these values in a computer (Stat:Fit) software. Therefore, identify the probability distribution that arrived in the local customers; and with that distribution, we immediately proceeded to perform the simulation process to obtain more information. It should be noted that this was done using the Monte Carlo method.

8.2.3 Stat:Fit

It is a discrete event simulation technology is used for is, design and improve manufacturing systems, logistics and other types of new systems or existents. Also, it allows you to accurately represent real-world processes, including their inherent variability and interdependencies to carry out predictive analysis of changes subsidies according to the environment and their key performance indicators [18–22].

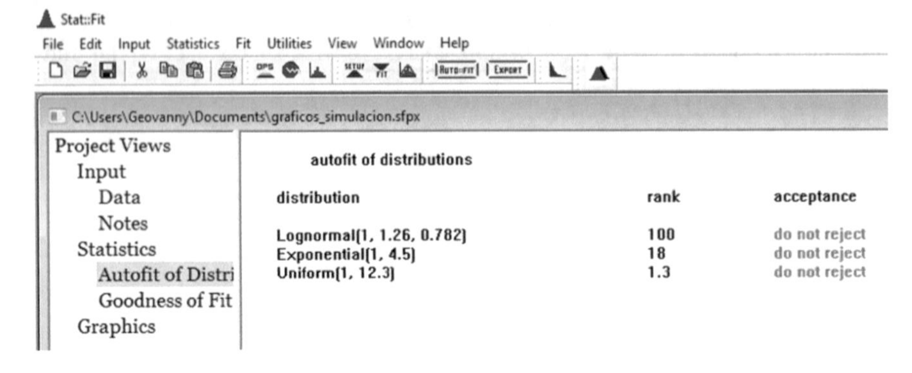

Fig. 8.1 Stat:Fit. Probability representations

Table 8.1 Total number of people who entered the store in 2 h

Day	Time (h)	People (Days)
1	2	17
2	2	24
3	2	21
4	2	22
5	2	26

Table 8.2 Daily demand for people

1	2	3	4	5	6	People (Days)
4	6	4	2	0	1	17
7	9	4	2	1	1	24
5	8	4	3	1	0	21
8	7	1	3	1	2	22
11	7	4	1	2	1	26

Stat:Fit is an essential tool because it helped us to obtain the correct acquisition of the probability distributions that will be used to perform the simulation [18, 20]. Table 8.1 shows the data obtained in the 5 days (Fig. 8.1).

Table 8.2 shows the demand for the product per person. In other words, the total of time which could be observed people in the establishment the purchase perishable (bananas).

8.3 Results and Discussion

The results obtained from the simulation of the sale of a perishable product were evaluated, by using the distribution of exponential probability. It was possible to get an approximation of the time and the quantity of the daily sale of bananas which are shown in Table 8.3.

Table 8.3 Values obtained with the Montecarlo simulation algorithm and exponential

People	Random	Exponential	Time elapsed	Bananas per person
1	0.07746816	0.46819385	0.46819385	1
2	0.23803562	1.57851563	2.04670948	2
3	0.06706321	0.40307128	2.44978076	1
4	0.83529049	10.472353	12.9221338	4
5	0.20217848	1.31150534	14.2336391	2
6	0.77430505	8.6433152	22.8769543	4
7	0.97705376	21.9170395	44.7939939	6
8	0.79892236	9.31392103	54.1079149	4
9	0.48047233	3.8022689	57.9101838	3
10	0.68185514	6.64982984	64.5600137	4
11	0.27710855	1.88417152	66.4441852	2
12	0.08064276	0.48820938	66.9323946	1
13	0.33965683	2.40965201	69.3420466	2
14	0.24126452	1.60317335	70.9452199	2
15	0.43799174	3.3459023	74.2911222	3
16	0.33065921	2.33106937	76.6221916	2
17	0.64160943	5.95818529	82.5803769	3
18	0.42940086	3.25781622	85.8381931	3
19	0.8416003	10.6991633	96.5373564	4
20	0.46578079	3.64034919	100.177706	3
21	0.44840288	3.45447498	103.632181	3
22	0.6225459	5.6572624	109.289443	3
23	0.67158902	6.46542266	115.754866	4
24	0.99754316	34.8902578	150.645124	6
25	0.03793025	0.22452576	150.869649	1
26	0.54952043	4.63031163	155.499961	3
27	0.05399549	0.32230418	155.822265	1
28	0.6103335	5.47237184	161.294637	3
29	0.16952991	1.07862602	162.373263	2
30	0.49480933	3.96475755	166.338021	3
31	0.44786513	3.44881701	169.786838	3
32	0.9895156	26.4650347	196.251872	6
33	0.59961057	5.31474749	201.56662	3
34	0.75168651	8.08875449	209.655374	4
35	0.28587615	1.95502575	211.6104	2
36	0.89955479	13.3440553	224.954455	5
37	0.84727621	10.9110443	235.8655	4
38	0.32533997	2.28510801	238.150608	2
39	0.8649646	11.6257839	249.776392	5
40	0.51214296	4.16748112	253.943873	3

(continued)

Table 8.3 (continued)

People	Random	Exponential	Time Elapsed	Bananas per person
41	0.20995116	1.36835138	255.312224	2
42	0.88599676	12.6088746	267.921099	5
43	0.48694932	3.87511349	271.796212	3
44	0.98432356	24.1292695	295.925482	6
45	0.02900505	0.17090717	296.096389	1
46	0.97204798	20.7712218	316.867611	6
47	0.84915698	10.9829936	327.850604	4
48	0.28661662	1.96104953	329.811654	2
49	0.20537782	1.33483656	331.14649	2
50	0.44317383	3.39968998	334.54618	3
51	0.68504703	6.70837909	341.254559	4
52	0.28114372	1.91667389	343.171233	2
53	0.96520472	19.4996521	362.670885	6
Sales Total				168

Table 8.4 Daily demand, reorder point and banana delivery time

Initial inventory	Daily demand	Final inventory	Order	Random	Delivery time
300	190	110	1	0.015328944	1
410	173	237	1	0.771438122	2
237	203	34	1	0.738401115	
334	181	153	1	0.069140553	1
453	192	261	1	0.569536984	2
261	213	48	1	0.028003454	
348	190	158	1	0.461136758	1
458	211	247	1	0.673638701	2
247	191	56	1	0.687927306	
356	196	160	1	0.127178192	1
460	189	271	1	0.050575435	1

Once obtained the daily store sales demand, proceeded to calculate the optimal quantity of order of business (EOQ). The initial inventory of bananas during the week was 300 units that will vary depending on the demand which is generated during the day. Furthermore, the reorder point is defined and determined a value below 200 units in the ending inventory of the day; which allowed to simulate a model of maximal and minimal (see Table 8.4).

It must be considered that the time it takes for the order made to the supplier to arrive will depend on the random number generated by the algorithm, where if the random number is ≤ 0.5 it will take one day for the product to arrive, whereas, if the random one is >0.5 and ≤ 1 the product will arrive in a maximum of two days.

It is demonstrated that the management model of maximum and minimum inventories if it can be applied to our research, since it allows us to have a maximum stock at each replenishment point, as long as the supplier complies with the agreed upon delivery date to business.

8.4 Conclusion

After carrying out the corresponding processes and using the methods of distribution and modeling mentioned above, it is concluded that the store or mini-market has a small loss in the estimated product (banana). Therefore, it is suggested that the orders be applied according to what was simulated in this work, to reduce the losses of products, which if this is controlled; could generate more savings in these products that will not be sold, for the reason that they will lose over time.

8.5 Future Work

This study can use a queue model, which will allow us to realize if the establishment has enough servers to supply the demand, and also, obtain a greater profit.

Finally, the implementation stage of this design can be done in a GUI with a high-level programming language which is aimed at solving the problems of this magnitude.

References

1. Rodríguez, Edwin Causado. 2015. Modelo de inventarios para control económico de pedidos en empresa comercializadora de alimentos. *Revista de Ingenierías: Universidad de Medellín* 14 (27): 15–15.
2. Escobar, John Willmer, Rodrigo Linfati, and Wilson Adarme Jaimes. 2017. Gestión de inventarios para distribuidores de productos perecederos. *Ingeniería y Desarrollo* 35 (1): 219–239.
3. Jara-Cordero, Sergio, Diana Sánchez-Partida, and José Luis Martínez-Flores. 2017. Análisis para la mejora en el manejo de inventarios de una comercializadora. *Septiembre* 1 (1): 1–18.
4. Rodríguez López, Manuel Guillermo, Flor Salazar Vázquez, and Jorge González Urgiles. 2018. *Control de inventarios con ajuste dinámico del punto de reorden - Un caso de estudio para empresas con productos perecibles y no perecibles, usando técnicas computacionales* 23: 13–20.
5. Osorio, Carlos Andrés. 2013. Modelos para el control de inventarios en las pymes. *Panorama* 2 (6).
6. Valdivia, Mudarra, Cicely Jobana, Zavaleta Contreras, and Santa Fania. 2018. El control interno de inventarios y su relación con la rentabilidad de la empresa minimarket san marcos SAC, periodo 2016.
7. Gutiérrez, Valentina, and Carlos Julio Vidal. 2014. Modelos de gestión de inventarios en cadenas de abastecimiento: revisión de la literatura. *Revista Facultad de Ingeniería* 43: 134–149.

8. Peña, Omaira, and Rafael Da Silva Oliveira. 2016. Factores incidentes sobre la gestión de sistemas de inventario en organizaciones venezolanas. *Telos: Revista de Estudios Interdisciplinarios en Ciencias Sociales* 18 (2): 187–207.
9. William, Richard, and Lopez Prado. 2018. El control de inventario como estrategia para el logro de rentabilidad en las mypes comerciales de la actividad ferretera ubicada en la comunidad urbana autogestionaria de huaycán distrito de ate-lima, periodo 2016.
10. Cubas García, Marleny Janet. 2016. El control de inventarios y su incidencia en la rentabilidad de la empresa artceramics imagen SAC, 2015.
11. Molina, Dolores. 2015. Gestión de inventarios: una herramienta útil para mejorar la rentabilidad.
12. Landeta, Juan Manuel Izar, Carmen Berenice Ynzunza Cortés, and Orlando Guarneros García. 2016. Variabilidad de la demanda del tiempo de entrega, existencias de seguridad y costo del inventario. *Contaduría y administración* 61 (3): 499–513.
13. Eduardo, Gutiérrez-González, Panteleeva Olga Vladimirovna, Hurtado-Ortiz Moisés Fernando, and González-Navarrete Carlos. 2013. Aplicación de un modelo de inventario con revisión periódica para la fabricación de transformadores de distribución. *Ingeniería, investigación y tecnología* 14 (4): 537–551.
14. García, Jesús Fernando Isaac, Sara Oranday Dávila, et al. 2012. Modelo probabilístico de quiebra de la pequeña y mediana empresa española. evidencia empírica. un modelo econométrico. *Contribuciones a la Economía* 67.
15. Ruiz Torres, Alex Jesús, José Humberto Ablanedo Rosas, and Jorge Ayala Cruz. 2012. Modelo de asignación de compras a proveedores considerando su flexibilidad y probabilidad de incumplimiento en la entrega. *Estudios gerenciales* 28 (122).
16. Garza, Juvencio Jaramillo, Jesús Fernando Isaac García, et al. 2014. Modelo probabilístico para medir, pronosticar, y prevenir la quiebra de las empresas pyme en nuevo león méxico. una herramienta para la planeación financiera y la toma de decisiones empresariales con evidencia empírica. *Observatorio de la Economía Latinoamericana* 195.
17. Jaramillo, Juvencio, G. Jesús Fernando Isaac, et al. 2015. Determinantes de la quiebra empresarial pyme en zacatecas. desarrollo de un modelo probabilístico-predictivo de la quiebra pyme. *Observatorio de la Economía Latinoamericana* 213: 1–24.
18. Benneyan, James C. 1998. Software review: Stat: Fit. *OR/MS Today* 25 (1): 38–41.
19. Klingstam, Pär, and Per Gullander. 1999. Overview of simulation tools for computer-aided production engineering. *Computers in Industry* 38 (2): 173–186.
20. Leemis, Lawrence M. 2002. Software review: Stat: Fit fitting continuous and discrete distributions to data. *OR/MS Today* 29 (3): 52–55.
21. Chi, Rosa Imelda Garcia, Arturo Eguia Alvarez, Gloria Emilia Izaguirre Cardenas, et al. 2015. Uso de la herramienta de software promodel como estrategia didáctica en el aprendizaje basado en competencias de simulación de procesos y servicios. *TECTZAPIC* 1.
22. Cabanillas, Alberto Cossa. 2012. Modelo de simulación para programar y controlar los recursos en una gestión hospitalaria. *Interfases* 5: 83–100.

Printed in the United States
By Bookmasters